T0211040

Mathematical Physics Studies

More information about this series at http://www.springer.com/series/6316

Giuseppe Gaeta · Miguel A. Rodríguez

Lectures on Hyperhamiltonian Dynamics and Physical Applications

 Springer

Giuseppe Gaeta
Dipartimento di Matematica
Università degli Studi di Milano
Milan
Italy

Miguel A. Rodríguez
Departamento de Física Teórica II,
Facultad de Ciencias Físicas
Universidad Complutense de Madrid
Madrid
Spain

ISSN 0921-3767
Mathematical Physics Studies
ISBN 978-3-319-85378-9
DOI 10.1007/978-3-319-54358-1

ISSN 2352-3905 (electronic)

ISBN 978-3-319-54358-1 (eBook)

Printed on acid-free paper

This Springer imprint is published by Springer Nature
The registered company is Springer International Publishing AG
The registered company address is: Gewerbestrasse 11, 6330 Cham, Switzerland

Hamilton's genius was recognised early.
[...]
In an unprecedented move, he was offered
a full professorship in Dublin
while still an undergraduate.
[...]
Unfortunately, the later years of
Hamilton's life were not happy ones.
The woman he loved married another
and he spent much time depressed,
mired in drink, bad poetry and quaternions.

(D. Tong; *Classical Dynamics*)

Acknowledgements

This book originates from our collaborative work over the years; we gratefully acknowledge support and hospitality in this time from Università degli Studi di Milano and Universidad Complutense de Madrid. Writing of this book was started while one of us (GG) was on sabbatical leave; he would like to thank his University for this opportunity, and SMRI for hospitality. The work of MAR was partially supported by Spain's MINECO under Grant No. FIS2015-63966-P, and by the Universidad Complutense de Madrid and Banco Santander under Grant No. GR3/14-910556.

Milan, Italy   Giuseppe Gaeta

Madrid, Spain   Miguel A. Rodríguez

Contents

Introduction

Hyperhamiltonian dynamics was formulated [60] as an attempt to generalize standard Hamiltonian dynamics from the symplectic to the hypersymplectic framework. Quite remarkably, it turns out that the generalization first considered, and based on symplectic geometry considerations, is also natural from the point of view of generalizing the description of Hamilton dynamics (on Kahler manifolds) from the point of view of complex (to hypercomplex) Analysis [120].

From the geometric point of view, hyperhamiltonian dynamics is based on hyperkahler geometry [13, 14]; this is known to be relevant in different physical contexts [14, 37, 40, 41, 49, 92, 126, 141, 150]. Remarkably, the HKLR theorem [92] guarantees that a quotient procedure exists for hyperkahler manifolds; in particular, this allows to build nontrivial hyperkahler manifolds from trivial ($\mathbf{R}^{4n}$ with Euclidean metric) ones; the Taub-NUT spaces are examples of hyperkahler manifolds admitting such a description.

The original physical motivation behind the construction of hyperhamiltonian dynamics lies in the desire to formulate a dynamics specifically adapted to systems with *spin*. These are primarily described by the Pauli and Dirac equations; in this sense, our work was successful, as both of these are well described within the hyperhamiltonian framework [66]. On the other hand, as already mentioned, hyperkahler geometry is also relevant in other fields of Physics, so that we expect our construction is also relevant there. As an application to these other physical problems, we point out the explicit identification of the hyperkahler structure in the Taub-NUT case [67].

Actually, in all of our discussion—in original papers and also in this book—we have put special emphasis on the task of obtaining (as far as possible) explicit, and possibly simple, results. Thus, some of our findings will actually reproduce results which were already known in geometry in abstract terms, but which are here given in concrete ones; e.g., the classification of hyperkahler maps in Chap. 4 is given in terms of explicit finite-dimensional matrices. Needless to say, we are aware that this will appear as a strong point to some readers, and as a weak one to others (or possibly to the same ones in different circumstances).

As said above, the HKLR theorem and quotient procedure allow to obtain nontrivial hyperkahler manifolds from Euclidean ones. Albeit manifolds obtained as quotient of Euclidean ones are not the most general type of hyperkahler manifolds, it turns out that the vast majority of physically relevant ones can be obtained in this way. This is one of the motivations for our insisting on detailed study of (several aspects of) Euclidean hyperkahler manifolds and the natural "standard" structures (see Chap. 1) defined on these.

The other reason for this insistence on the Euclidean case is of course that it is much simpler than the general one, and one can thus obtain more detailed and explicit results.

In the general case, one has in particular several complications due to the presence of a nontrivial holonomy group [5, 25, 34, 106], which in the case of a 4n-dimensional hyperkahler manifold is generically Sp(n) [21]. The holonomy of hyperkahler manifolds is discussed in several excellent recent publications (see, e.g., [25, 100, 104, 133]), to which the reader is referred. Here, we give a very sketchy discussion, and more references, in Appendix A; and assume the reader has some basic knowledge of it and its role in some physical problems (e.g., at the level of [123]).

The text is organized in the following way. In Chap. 1, we provide some background on hyperkahler structures and manifolds—and for the sake of fixing notation, also of standard Hamiltonian dynamics and symplectic geometry. In Chap. 2, we provide the basic definitions of hyperhamiltonian dynamics and introduce a tool which will be of special use in the following, i.e., the *hyperkahler form*. This is essential in defining transformations which preserve (all or certain aspects of) the hyperkahler structure and which generalize the familiar canonical transformations of Hamiltonian dynamics and symplectic geometry; these are the *canonical* and *hyperkahler* maps, discussed in detail, respectively, in Chaps. 3 and 4. In particular, in Chap. 3, we will show that—in the same way as any Hamiltonian vector field is the generator of a one-parameter group of canonical transformations for the underlying symplectic structure—any hyperhamiltonian vector field is the generator of a one-parameter group of canonical transformations for the underlying hyperkahler structure. While in the standard Hamiltonian case, the Hamiltonian vector fields are the only ones with this property, in the hyperhamiltonian case, the property is shared by vector fields which are associated with *dual* hyperkahler structure (in the sense defined in Chap. 1), i.e., *Dirac vector fields*. In Chap. 5, we focus our attention on *integrable* hyperhamiltonian systems, and in Chap. 6, we consider *perturbations* of these in terms of (a suitable generalization of) the Poincaré-Birkhoff normal forms theory. Finally, in Chap. 7, we consider several physical applications, and in particular those mentioned above: the Pauli and Dirac equations, and Taub-NUT manifolds.

This book is completed by three short appendices: Appendix A deals, as already mentioned, with *holonomy*, while Appendix B discusses the variational approach to hyperhamiltonian dynamics. The third appendix, i.e., Appendix C, is of a different character and gives a sketchy account of works concerning the quaternion approach to quantum mechanics; this is quite different from our field of interest, which only concerns classical mechanics, but the readers will be interested in having some reference about works in that direction (we thank a reviewer for stressing the appropriateness of such a discussion).

Chapters 1 and 2 are needed for the following ones, which are reasonably independent of each others, except for Chap. 6 which relies on Chap. 5. To obtain that independence, we have some duplication in our discussion.

In the text, we have adopted a simplified spelling for the *Kahler* (rather than Kähler) and *hyperkahler* (rather than hyperkähler or hyperKähler) terms.

Chapter 1
Background Material

1.1 Hamiltonian Dynamics

In this section we will discuss some basic facts and background notions. We will start by recalling some well known definitions in Hamiltonian dynamics [10, 12, 30, 109] and in Kahler geometry [18, 30, 95, 106, 121, 123, 133] (which will be of use in view of their generalization later on), and will then pass to collect some basic notions on Kahler and hyperkahler manifolds and structures [13, 14, 74, 92, 100, 104, 115, 133]. Finally we will restrict to the special (but relevant!) case of Euclidean such manifolds, recalling that in this case there are some "standard" hyperkahler structures.

1.1.1 Symplectic Forms and Manifolds

Let us consider a smooth[1] m-dimensional manifold M.

A *symplectic form* $\omega \in \Lambda^2(M)$ on M is a closed, non-degenerate two form; it can be easily seen that the requirement of non-degeneracy implies that symplectic forms can only exist on manifolds of even dimension, i.e. $m = 2n$. We also say that ω defines a *symplectic structure*[2] on M.

A smooth manifold equipped with a symplectic form (equivalently, with a symplectic structure) is said to be a *symplectic manifold*.

If ω admits a (in general, non unique) form α such that $\omega = d\alpha$, we also say that α is a "symplectic potential" for ω. Note that if $M = \mathbf{R}^{2n}$, and more generally if

---

[1]By "smooth" we will always mean C^∞; note also that all objects (manifolds, vector fields, forms, etc.) we consider are *real* unless explicitly stated otherwise.

[2]Many excellent discussions of Symplectic Geometry are available in the literature, see e.g. [10, 30, 82]; the reader is referred to these for details.

© Springer International Publishing AG 2017
G. Gaeta and M.A. Rodríguez, *Lectures on Hyperhamiltonian Dynamics and Physical Applications*, Mathematical Physics Studies, DOI 10.1007/978-3-319-54358-1_1

the second cohomology class of M vanishes, ω is not only closed, but exact and any symplectic form admits a symplectic potential.

Note that a symplectic manifold does *not* necessarily have a metric; in the case a Riemannian metric g is defined on a symplectic manifold, there are no constraints on the relations between the metric g and the symplectic form ω.

We will be interested in cases where a metric is actually defined on the manifold under consideration; thus we will assume a Riemannian metric g is defined in M, and hence on local charts $U \subseteq M$. For the time being (i.e. until next section), this has no definite relation with ω.

If there are local coordinates x^i on $U \subseteq M$, then ω is written locally as

$$\omega = \frac{1}{2} A_{ij} \, dx^i \wedge dx^j \, , \tag{1.1}$$

with $A = A(x)$ a smooth skew-symmetric matrix function; the non-degeneration of the form ω is reflected into the non-degeneration of the matrix $A(x)$ at all $x \in U$ (we will always understand, unless explicitly noted, sum over repeated upper and lower indices).

The well known Darboux theorem states that there are always local coordinates (p, q), with p and q n-dimensional, such that in these we get $\omega = dp_i \wedge dq^i$; in the notation used above, this means there are local coordinates x^i ($i = 1, \ldots, 2n$) such that in these we have

$$A = \begin{pmatrix} 0 & -I_n \\ I_n & 0 \end{pmatrix} \, ,$$

with I_n the n-dimensional identity matrix. (One should think of these as $x^i = p_i$ for $i = 1, \ldots, n$ and $x^i = q^{i-n}$ for $i = n+1, \ldots, 2n$.) As a consequence of the Darboux theorem, all symplectic manifolds of fixed dimension $2n$ are isomorphic (locally).

As discussed e.g. in [10], the Darboux coordinates can be used to define (locally) a *symplectic metric*; we stress once again that if a metric g is defined in M, the symplectic metric does not necessarily (and in general does not) agree with this.

1.1.2 Hamiltonian Dynamics and Hamiltonian Vector Fields

Let us consider the ring of smooth real functions on M. To each such function H, we can associate a Hamiltonian vector field[3] X_H by

$$X_H \lrcorner \omega = dH \, . \tag{1.2}$$

Note that this is well defined and unique, due to the non-degeneration of ω.

---

[3] Vector fields are always assumed to be smooth.

1.1.2.1 Properties of Hamiltonian Vector Fields

It should be stressed that, for any H, X_H does necessarily preserve ω, i.e. (with $\mathcal{L}$ the Lie derivative)

$$\mathcal{L}_{X_H}(\omega) \ = \ 0 \tag{1.3}$$

In fact, for a generic vector field X and form ω, we have by Cartan's formula

$$\mathcal{L}_X(\omega) \ = \ X \lrcorner (\mathrm{d}\omega) \ + \ \mathrm{d}(X \lrcorner \omega) \ ;$$

the symplectic form is closed, so that $\mathrm{d}\omega = 0$, and we have then using (1.2)

$$\mathcal{L}_{X_H}(\omega) \ = \ \mathrm{d}(\mathrm{d}H) \ = \ 0 \ .$$

The vector fields satisfying $\mathcal{L}_X(\omega) = 0$ are called symplectic vector fields. Then, any Hamiltonian vector field is symplectic, as stated by (1.3). Note that the converse statement is not always true.

If we have local coordinates x^i on $U \subseteq M$, the vector field X_H will read as

$$X_H \ = \ f^i(x) \, \frac{\partial}{\partial x^i} \ , \tag{1.4}$$

and the components f^i are given by[4]

$$f^i \ = \ A^{ij} \, \frac{\partial H}{\partial x^j} \ , \quad f \ := \ A \, \nabla H \ , \quad A^{ij} A_{jk} = \delta^i_k \ . \tag{1.5}$$

Let us consider the case where M is endowed with a Riemannian metric g and with a volume form Ω. It is easily seen that in this case necessarily X_H preserves Ω as well; we also say that Hamiltonian vector fields are *Liouville*.

Remark 1.1 We recall that a Liouville vector field is one which preserves the volume in phase space, i.e. with zero divergence. We take this occasion to mention that usually by *volume* of $B \subseteq M$ we mean the *absolute value* of the integral of the volume form over B; thus an orthogonal orientation-reversing map will change the volume form Ω into $-\Omega$ but will leave the (absolute value of the) volume unchanged;

---

[4]Note that A_{ij} are the elements of the matrix A (defined in equation (1)) while A^{ij} are the elements of the inverse matrix A^{-1}. If we think of A as a (symplectic) metric this notation is quite natural. It should be stressed that here we have used the notation with upper and lower indices—in order to prepare for the situation where a Riemannian metric *is* defined on M—but we do *not* need assuming such a metric is defined. In fact, as mentioned above, a symplectic manifold does not require the introduction of a metric. Note however that we have assumed some metric g is defined in M and hence in $U \subseteq M$; we can thus dispense with such worries and just use this to raise and lower indices when needed.

see the discussion in Sect. 1.5. Note that for (continuous families of non-singular maps generated by) vector fields such problems do not arise, as these cannot reverse orientation.   ⊙

1.1.2.2   The Lie Algebra Structure of Hamiltonian Vector Fields

Let us now consider the commutator of Hamiltonian vector fields; if we consider the vector fields X_1 and X_2 associated to the Hamiltonians H_1 and H_2, this will be a vector field

$$Y := [X_1, X_2] \tag{1.6}$$

given in coordinates by

$$Y = (f^j \partial_j g^i - g^j \partial_j f^i) \, \partial_i := h^i \partial_i , \tag{1.7}$$

where we have written

$$X_1 = f^i(x) \, \partial_i , \quad X_2 = g^i(x) \, \partial_i .$$

It is easily checked that the vector field Y is itself Hamiltonian.

This can be shown[5] by working in coordinates: in fact, using integration by parts,

$$
\begin{aligned}
h^i &= (A^{jm} \partial_m H_1) \, \partial_j [A^{i\ell} \partial_\ell H_2] - (A^{jm} \partial_m H_2) \, \partial_j [A^{i\ell} \partial_\ell H_1] \\
&= A^{jk} \frac{\partial}{\partial x^j} \left(A^{\ell i} \left(\frac{\partial H_1}{\partial x^\ell} \frac{\partial H_2}{\partial x^k} - \frac{\partial H_1}{\partial x^k} \frac{\partial H_2}{\partial x^\ell} \right) \right) \\
&= A^{ji} \frac{\partial}{\partial x^j} \left(A^{\ell k} \frac{\partial H_1}{\partial x^\ell} \frac{\partial H_2}{\partial x^k} \right)
\end{aligned}
$$

where we have also used the closeness property of the 2-form ω:

$$\partial_i A_{jk} + \partial_j A_{ki} + \partial_k A_{ij} = 0 \tag{1.8}$$

$$A^{\ell k} \frac{\partial A^{mj}}{\partial x_k} + A^{mk} \frac{\partial A^{j\ell}}{\partial x_k} + A^{jk} \frac{\partial A^{\ell m}}{\partial x_k} = 0 \tag{1.9}$$

To summarize, the symplectic vector fields on a symplectic manifold form a Lie algebra, while the Hamiltonian vector fields are an ideal of this Lie algebra.

---

[5]We are showing the proof of this well known fact in order to emphasize, later on, the differences with the hyperhamiltonian case.

1.1.2.3 The Lie Algebra Structure of the Function Ring $C^\infty(M)$

We can use the symplectic form ω to introduce a Lie algebra structure in $C^\infty(M)$. The Poisson bracket is defined as:

$$\{H_1, H_2\} := \omega(X_1, X_2) = X_2(H_1) = -X_1(H_2), \quad H_1, H_2 \in C^\infty(M)$$

where X_1 and X_2 are the corresponding Hamiltonian vector fields. In coordinates, it can be written as:

$$\{H_1, H_2\} = A^{\ell k}\frac{\partial H_1}{\partial x^\ell}\frac{\partial H_2}{\partial x^k} \tag{1.10}$$

and has the properties of a Lie bracket: it is obviously skew-symmetric and the Jacobi identity (up to constant functions) is a consequence of the closeness property of the symplectic form ω, in fact, equivalent to the constraint (1.9). We can also prove that the Leibniz rule holds:

$$\{H_1 H_2, F\} = H_1\{H_2, F\} + H_2\{H_1, F\}$$

The relation between these two structures has been explained in the previous section where we proved the result:

$$[X_1, X_2] = -X_{\{H_1, H_2\}}$$

that is, there is a homomorphism (the minus sign can be avoided redefining the Hamiltonian vector field or the Poisson bracket) between the Lie algebra of Hamiltonian vector fields and the Poisson algebra of Hamiltonian functions. The same result could have also been obtained in a more intrinsic way. In fact, if H is a Hamiltonian function, the corresponding Hamiltonian vector field is defined by (1.2).

If X_1, X_2 are two Hamiltonian vector fields, using the Leibniz rule:

$$\mathcal{L}_{X_1}(X_2 \lrcorner \omega) = (\mathcal{L}_{X_1} X_2) \lrcorner \omega + X_2 \lrcorner (\mathcal{L}_{X_1}\omega) = [X_1, X_2] \lrcorner \omega;$$

because X_1 is symplectic. But we can also write, using the Cartan identity

$$\mathcal{L}_{X_1}(X_2 \lrcorner \omega) = d(X_1 \lrcorner (X_2 \lrcorner \omega)) + X_1 \lrcorner d(X_2 \lrcorner \omega) = d\omega(X_2, X_1);$$

The identity

$$[X_1, X_2] \lrcorner \omega = d\omega(X_2, X_1)$$

shows that $[X_1, X_2]$ is a Hamiltonian vector field and the corresponding Hamiltonian function is: $\omega(X_2, X_1) \equiv \{H_2, H_1\}$.

1.1.2.4   Poisson Tensor and Poisson Structure

Poisson brackets have been defined in a symplectic manifold using the properties of the symplectic form. We can introduce a Poisson bracket in an abstract way in any commutative algebra over the reals [31, 113, 148]. A bilinear map is a Poisson bracket if it is skewsymmetric, satisfies the Jacobi identity, and the Leibniz rule. In particular, given a manifold M, we can introduce a Poisson bracket in the algebra of smooth functions $C^\infty(M)$. If the manifold is endowed with a symplectic structure, the Poisson bracket derived from the symplectic form is a particular case of a Poisson structure.

Given a Poisson manifold (that is, a manifold with a Poisson structure $\{\cdot, \cdot\}$, in its function space), there exist a (unique) contravariant skewsymmetric 2-tensor, Π, satisfying

$$\Pi(\mathrm{d}f, \mathrm{d}g) := \{f, g\}, \quad f, g \in C^\infty(M) . \tag{1.11}$$

The Jacobi identity for the Poisson bracket is then equivalent to the vanishing of the Schouten bracket for the Poisson tensor:

$$[\Pi, \Pi] = 0 \tag{1.12}$$

a sort of integrability condition. This can be written in coordinates as:

$$\Pi^{ij} \frac{\partial \Pi^{k\ell}}{\partial x^i} + \Pi^{ik} \frac{\partial \Pi^{\ell j}}{\partial x^i} + \Pi^{i\ell} \frac{\partial \Pi^{jk}}{\partial x^i} = 0 ,$$

which is also the same expression (1.9) satisfied by A^{-1} in the case of a symplectic manifold. The Poisson bracket has the same expression as in (1.10).

Just as in a symplectic manifold, we can introduce Hamiltonian vector fields associated to functions in $C^\infty(M)$ using the Poisson bracket (the derivation property being a consequence of the Leibniz rule satisfied by the Poisson bracket).

The rank of a Poisson manifold is defined as the rank of the Poisson tensor at a point $x \in M$. It turns out that, a Poisson manifold with constant rank is foliated by symplectic leaves (with the same dimension). The symplectic structure is induced from the Poisson structure. Locally, a Poisson manifold is the product of a symplectic manifold and a Poisson manifold with rank equal to zero at some point. Symplectic manifolds and the dual of Lie algebras are the most immediate examples of Poisson manifolds.

Although we will no insist here on this point, we would like to mention that Poisson manifolds are related to the theory of algebroids and grupoids. To be more precise, the cotangent bundle of a Poisson manifold is a Lie algebroid [148].

1.1.2.5 The Lie Algebra Structure of Constants of Motions

If (M, ω) is a symplectic manifold and H a Hamiltonian, with Hamiltonian vector field X, the flow φ_t (which is a symplectic transformation, that is, leaves ω invariant) also leaves H invariant:

$$\varphi_t H = H \qquad (1.13)$$

If $F \in C^\infty(M)$, the infinitesimal action of the flow X on F is given by

$$X(F) = \{F, H\}$$

and the quantity F is a constant of the motion if and only if

$$\{H, F\} = 0$$

We remark that if F is invariant under the flow generated by H, and then $\{H, F\} = 0$, the quantity H is invariant under the flow generated by F. If F, G are two invariant quantities with respect to the Hamiltonian H, using Jacobi identity, we can write

$$\{H, \{F, G\}\} = \{F, \{H, G\}\} - \{G, \{H, F\}\}$$

and then, the Poisson bracket of two constants of motion is another constant of motion. The set of constants of motion of a Hamiltonian has a structure of Lie algebra. Note that any two constants of motion do not necessarily commute.

1.2 Integrable Hamiltonian Systems

Integrable systems provide the best studied models in Physics. In finite or infinite dimension, integrable systems, which are rare, are also widely used. As we have discussed in the previous section, the constants of the motion relative to a Hamiltonian H form a Lie subalgebra of $C^\infty(M)$ with H as a central element. In some cases this subalgebra is Abelian and maximal. A Hamiltonian system is called integrable if there exist n globally defined (in a dense open subset of M) constants of motion (first integrals), including the Hamiltonian, which are in involution,

$$\{F_i, F_j\} = 0,$$

and are functionally independent (that is, the linear subspace of 1-forms generated by dF_i has dimension n). We also assume that the corresponding Hamiltonian vector fields are complete (the flows are defined for all t).

Note that, since the constants of motion F_i commute, the corresponding flows also commute and we have an action of $\mathbf{R}^n$ over the manifold M. if $F = (F_1, \ldots, F_n)$, the level sets $F^{-1}(c)$, with $c \in \mathbf{R}$, are preserved under this action.

The theorem of Arnold-Liouville describes the orbits of an integrable Hamiltonian system (see for instance [10, 12, 30, 81, 85]):

Theorem 1.1 (Arnold-Liouville) *If* $f : M \to \mathbf{R}^n$ *is an integrable system with compact connected fibers, the flows of the Hamiltonian vector fields* X_i *induce a diffeomorphism* $F^{-1}(c) \sim \mathbf{R}^n/L_c$ *for all* $c \in R$, *where* L_c *is a discrete subgroup of the additive group* $\mathbf{R}^n$. *The discrete subgroup* L_c *is called the* period lattice *and the level sets* $F^{-1}(c)$ *the* Liouville tori.

1.3 Kahler Manifolds

An almost complex manifold is a $2n$ dimensional real manifold endowed with a smooth field of complex structures on TM (that is, linear endomorphisms such that $J_x^2 = -1$ at each tangent space $T_x M$). The necessary and sufficient condition for an almost complex manifold to become a complex manifold (a $2n$-dimensional real manifold with a holomorphic atlas) is provided by the Newlander-Nirenberg theorem [13, 123, 125]:

Theorem 1.2 (Newlander-Nirenberg) *An almost complex structure is a complex structure if and only if the Nijenhuis tensor (the torsion) of any two vector field vanishes:*

$$N(X, Y) = 2([JX, JY] - [X, Y] - J[X, JY] - J[JX, Y]) = 0$$

A complex manifold (complex structure J) with a compatible Riemannian metric, g [that is, the complex structure is orthogonal, $g(JX, JY) = g(X, Y)$] is called a *Hermitian manifold*. In such a manifold we can define a 2-form ω

$$\omega(X, Y) := g(X, JY), \tag{1.14}$$

the associated Kahler form. The metric is called a Kahler metric, and when ω is closed (that is, when we have a symplectic form) we will say (M, J, g) is a Kahler manifold [29].[6]

The conditions for a complex Riemannian manifold to be a Kahler manifold are identified by the following theorem [28, 123].

Theorem 1.3 (Calabi) *Let M be a complex manifold with a compatible metric g and a Levi-Civita connection ∇. Then the following statements are equivalent: (i) g is a Kahler metric; (ii)* $d\omega = 0$; *(iii)* $\nabla J = 0$.

---

[6]It should be stressed that not all symplectic manifolds admit a Kahler structure; see e.g. [116].

As a consequence of this result, if a manifold is provided with an almost complex structure J and a compatible Riemann metric g, such that J is parallel with respect to the Levi-Civita connection ($\nabla J = 0$), then J is a complex structure (by the Newlander-Nirenberg theorem) and the manifold is Kahler.

1.4 Hyperkahler Manifolds and Structures

1.4.1 Hyperkahler Structures

Having recalled some very basic notion in the standard Hamiltonian case, we can pass to consider background notions needed for the generalization of Hamiltonian Dynamics to be considered in the following, in particular those related to *hyperkahler* structures and geometry [13, 14, 74, 87, 88, 92, 100, 104, 115, 133, 136].

Definition 1.1 A *hyperkahler manifold* is a real smooth orientable Riemannian manifold (M, g), in which we choose an orientation, of dimension $m = 4n$ equipped with three almost-complex structures J_1, J_2, J_3 which:

(i) are covariantly constant under the Levi-Civita connection, $\nabla J_\alpha = 0$ [hence they are actually complex structures on (M, g)]; and
(ii) satisfy the quaternionic relations, i.e.

$$J_\alpha J_\beta = \varepsilon_{\alpha\beta\gamma} J_\gamma - \delta_{\alpha\beta} I \qquad (1.15)$$

with $\varepsilon_{\alpha\beta\gamma}$ the completely antisymmetric (Levi-Civita) tensor.

Simple examples of hyperkahler manifolds are provided by quaternionic vector spaces $\mathbf{H}^k$ and by the cotangent bundles of complex manifolds [92, 100].

Note that the relations (1.15) imply that the J_α satisfy the su(2) commutation relations, but also involve the multiplication structure.

We denote the ordered triple $\mathbf{J} = (J_1, J_2, J_3)$ as a *hyperkahler structure* on (M, g). We will denote a hyperkahler manifold as $(M, g; J_1, J_2, J_3)$, or simply as $(M, g; \mathbf{J})$.

Obviously a hyperkahler manifold is also Kahler with respect to any linear combination $J = \sum_\alpha c_\alpha J_\alpha$ such that $|c|^2 := c_1^2 + c_2^2 + c_3^2 = 1$; thus we have a S^2 sphere of Kahler structures on M. More precisely, we introduce the space

$$\mathbf{Q} := \Big\{ \sum_\alpha c_\alpha J_\alpha \, , \, c_\alpha \in \mathbf{R} \Big\} \approx \mathbf{R}^3 \, , \qquad (1.16)$$

also called the **Q**-*structure*[7] on (M, g) spanned by (J_1, J_2, J_3) [3]; and denote by $\mathbf{S} \approx S^2$ the unit sphere in this space. Points in **S** are in one to one correspondence with those Kahler structures on (M, g) which are in the linear span of the given basis structures J_α, and opposite points correspond to complex conjugate structures. The sphere **S** will play a central role in our discussion and deserves a special name.

Remark 1.2 Given a hyperkahler manifold M, this may be the product of (non-trivial) lower-dimensional hyperkahler manifolds $M_{(k)}$, which we will consider as "components" of M; in this case we say that M is a *reducible hyperkahler manifold*; if M has no non-trivial hyperkahler submanifolds, we say it is irreducible. For a reducible hyperkahler manifold, the hyperkahler structures on the $M_{(k)}$ components will be the restriction of the hyperkahler structures defined on M to $M_{(k)}$. In the decomposition of M in terms of components, we will always assume the latter are irreducible. A trivial example of reducible hyperkahler manifold is provided by $\mathbf{R}^{4n}$ with standard structures (see Sect. 1.4.3). $\qquad\qquad\qquad\qquad\qquad\qquad\qquad\odot$

Definition 1.2 The unit sphere **S** in **Q**, i.e. the set

$$\mathbf{S} := \Big\{ \sum_\alpha c_\alpha J_\alpha \, , \ c_\alpha \in \mathbf{R} \ : \ |c|^2 := \sum_\alpha c_\alpha^2 = 1 \Big\} \approx S^2 \subset \mathbf{R}^3 \, , \qquad (1.17)$$

is the *Kahler sphere* corresponding to the hyperkahler structure $\mathbf{J} = (J_1, J_2, J_3)$.

Definition 1.3 Two hyperkahler structures on (M, g) defining the same structure **Q**, and hence the same Kahler sphere **S**, are said to be *equivalent*. An equivalence class of hyperkahler structures is identified with the corresponding **Q**-structure, and viceversa.

There are obvious symplectic counterparts to the notions defined above, the correspondence being through the Kahler relation (1.14). Thus the symplectic forms ω_α correspond to the J_α, and $(M, g; \omega_1, \omega_2, \omega_3)$ is a *hypersymplectic manifold*. Any nonzero linear combination of the ω_α, i.e. any $\mu \neq 0$ in

$$\mathcal{Q} := \Big\{ \mu = \sum_\alpha c_\alpha \omega_\alpha \, , \ c_\alpha \in \mathbf{R} \Big\} \approx \mathbf{R}^3 \qquad\qquad (1.18)$$

is also a symplectic structure on M; in other words we have a punctured three dimensional space $\mathbf{R}^3 \backslash \{0\}$ of symplectic structures in M. Denote by $\mathcal{S}$ the unit sphere in $\mathcal{Q}$; the $\mu \in \mathcal{S}$ are *unimodular* symplectic structures in M. Obviously the sphere $\mathcal{S}$ corresponds to **S** via the Kahler relation; hence $\mathcal{S}$ is the *symplectic Kahler sphere* for the hyperkahler structure (J_1, J_2, J_3), and two hypersymplectic structures defining the same $\mathcal{S}$ are *equivalent*.

---

[7]In our earlier works, this was denoted as the *quaternionic structure*; however this notation is potentially confusing, as the notion of quaternionic structure is routinely used in the literature in a different sense [74, 115]; thus we prefer to use this (less suggestive but also less likely to cause misunderstandings) denomination.

Let us consider two equivalent structures $\mathbf{J}$ and $\widetilde{\mathbf{J}}$; by definition these generate the same three-dimensional linear space $\mathbf{Q}$, hence each of them can be written in term of the other. In particular we can write

$$\widetilde{J}_\alpha = R_{\alpha\beta} J_\beta \qquad (1.19)$$

(the metric in $\mathbf{Q}$ is Euclidean, so we will write both indices as lower ones for typographical convenience; note that sum over repeated indices is still implied, here and in the following). Now the requirement that $\widetilde{J}_\alpha^2 = J_\alpha^2 = -I$ forces the (real, three-dimensional) matrix R to be orthogonal, $R \in O(3)$. Moreover, the quaternionic relations (1.15) require $(\text{Det}(\widetilde{J}_\alpha) = \text{Det}(J_\alpha) = 1)$, hence also $(\text{Det}(R) = 1)$ in other words we must actually have $R \in SO(3)$.

The same argument also applies to equivalent hypersymplectic structures: in this case we also conclude that if $\{\omega_1, \omega_2, \omega_3\}$ and $\{\widetilde{\omega}_1, \widetilde{\omega}_2, \widetilde{\omega}_3\}$ are equivalent hypersymplectic structures, then necessarily $\widetilde{\omega}_\alpha = R_{\alpha\beta}\omega_\beta$, with $R \in SO(3)$.

Remark 1.3 In the case M is a linear space, $M = \mathbf{R}^{4n}$, it would be rather natural to expect that hyperkahler structures which are related by a linear transformation L are equivalent. However, the map L should preserve the Riemannian metric, i.e. be orthogonal. Moreover the hyperkahler structures related by such a linear transformation should define the same Kahler sphere. This means that L should also preserve the orientation of the ambient manifold; thus it will be not only orthogonal but special orthogonal (i.e. $\text{Det}(L) = +1$). Thus for $M = \mathbf{R}^{4n}$, the action of $SO(4n)$ will map the hyperkahler structure into an equivalent one.   ⊙

1.4.2 Hyperkahler Structures in Coordinates

The results we will discuss in these notes are of local nature, so we can work on a single chart of the hyperkahler manifold $(M, g; \mathbf{J})$. In the following we will use local coordinates x^i $(i = 1, \ldots, 4n)$; it will be useful to have a standard notation for expressing the objects introduced above in coordinates.

The metric g is defined in coordinates by $g_{ij}\, dx^i\, dx^j$ (we will use the same letter for its corresponding matrix); when using shorthand notation (with no indices) we will denote the contravariant metric tensor g^{ij} by g^{-1}.

The complex structures J_α and the associated Kahler symplectic forms ω_α will be written as

$$\begin{aligned} J_\alpha &= (Y_\alpha)^i{}_j\, \partial_i \otimes dx^j \\ \omega_\alpha &= \tfrac{1}{2}(K_\alpha)_{ij}\, dx^i \wedge dx^j\ ; \end{aligned} \qquad (1.20)$$

where the wedge product is defined as $dx^i \wedge dx^j = dx^i \otimes dx^j - dx^j \otimes dx^i$. We will also consider tensors of type $(2, 0)$ associated to these, i.e.

$$M_\alpha^{ij} = g^{i\ell} K_{\ell m}^\alpha g^{mj} . \tag{1.21}$$

Note that here M_α, Y_α, K_α are in general functions of the point x, and are of course not independent (we prefer to have distinct notations for the tensor fields Y_α, $K_\alpha = g Y_\alpha$, $M_\alpha = Y_\alpha g^{-1}$ as these will be useful in writing subsequent equations in compact form without the need to write down all the indexes; note $K_\alpha^{-1} = -M_\alpha$, and of course $Y_\alpha^{-1} = -Y_\alpha$).

The quaternionic relations (1.15) are reflected into the same relations being satisfied by the matrices Y_α, and similar ones—involving also g—by the K_α and M_α, i.e.

$$\begin{aligned}
Y_\alpha Y_\beta &= \varepsilon_{\alpha\beta\gamma} Y_\gamma - \delta_{\alpha\beta} I \\
K_\alpha g^{-1} K_\beta &= \varepsilon_{\alpha\beta\gamma} K_\gamma - \delta_{\alpha\beta} g \\
M_\alpha g M_\beta &= \varepsilon_{\alpha\beta\gamma} M_\gamma - \delta_{\alpha\beta} g^{-1} .
\end{aligned} \tag{1.22}$$

Similarly, the fact that the J_α are covariantly constant implies that $\nabla Y_\alpha = 0$ as well; as g is by definition also covariantly constant under its associated Levi-Civita connection, we also have $\nabla K_\alpha = 0$, $\nabla M_\alpha = 0$.

1.4.3   Standard Structures in $\mathbf{R}^4$ and $\mathbf{R}^{4n}$

In the following we will make reference to "standard" hyperkahler and hypersymplectic structures in $\mathbf{R}^{4n}$; these are obtained from standard structures in $\mathbf{R}^4$ (with Euclidean metric) [60]. We will consider the standard volume form $\Omega = dx^1 \wedge dx^2 \wedge dx^3 \wedge dx^4$ in $\mathbf{R}^4$.

There are two such standard structures, differing for their orientation. The positively-oriented standard hyperkahler structure is given by

$$Y_1 = \begin{pmatrix} 0 & 1 & 0 & 0 \\ -1 & 0 & 0 & 0 \\ 0 & 0 & 0 & 1 \\ 0 & 0 & -1 & 0 \end{pmatrix}, \quad Y_2 = \begin{pmatrix} 0 & 0 & 0 & 1 \\ 0 & 0 & 1 & 0 \\ 0 & -1 & 0 & 0 \\ -1 & 0 & 0 & 0 \end{pmatrix},$$

$$Y_3 = \begin{pmatrix} 0 & 0 & 1 & 0 \\ 0 & 0 & 0 & -1 \\ -1 & 0 & 0 & 0 \\ 0 & 1 & 0 & 0 \end{pmatrix} . \tag{1.23}$$

To these complex structures correspond the symplectic structures, satisfying $(1/2)(\omega_\alpha \wedge \omega_\alpha) = \Omega$ (no sum on α),

$$\begin{aligned}
\omega_1 &= dx^1 \wedge dx^2 + dx^3 \wedge dx^4, & \omega_2 &= dx^1 \wedge dx^4 + dx^2 \wedge dx^3, \\
\omega_3 &= dx^1 \wedge dx^3 + dx^4 \wedge dx^2.
\end{aligned} \tag{1.24}$$

The negatively-oriented standard hyperkahler structure is given by

$$\widehat{Y}_1 = \begin{pmatrix} 0 & 0 & 1 & 0 \\ 0 & 0 & 0 & 1 \\ -1 & 0 & 0 & 0 \\ 0 & -1 & 0 & 0 \end{pmatrix}, \quad \widehat{Y}_2 = \begin{pmatrix} 0 & 0 & 0 & -1 \\ 0 & 0 & 1 & 0 \\ 0 & -1 & 0 & 0 \\ 1 & 0 & 0 & 0 \end{pmatrix},$$

$$\widehat{Y}_3 = \begin{pmatrix} 0 & -1 & 0 & 0 \\ 1 & 0 & 0 & 0 \\ 0 & 0 & 0 & 1 \\ 0 & 0 & -1 & 0 \end{pmatrix}. \tag{1.25}$$

In this case, to these complex structures correspond the symplectic structures

$$\begin{aligned}
\widehat{\omega}_1 &= dx^1 \wedge dx^3 + dx^2 \wedge dx^4, & \widehat{\omega}_2 &= dx^4 \wedge dx^1 + dx^2 \wedge dx^3, \\
\widehat{\omega}_3 &= dx^2 \wedge dx^1 + dx^3 \wedge dx^4;
\end{aligned} \tag{1.26}$$

these satisfy $(1/2)(\omega_\alpha \wedge \omega_\alpha) = -\Omega$ (again with no sum on α).

Note that $[Y_\alpha, \widehat{Y}_\beta] = 0$ for all α, β. The existence of these two equivalent (and oppositely oriented) mutually commuting real representations of su(2) (and hence of the group SU(2) as well) is of course related to the quaternionic nature of SU(2) in the classification given by the real version of Schur Lemma (see e.g. Chap. 8 of [105], in particular Theorem 1.3 there).

Note also that while the su(2) commutation relations are satisfied by any representation, the condition $J_\alpha^2 = -I$ imply that the tensors J_α are represented, at any given point, by a sum of copies of the two (oppositely oriented) fundamental representations, i.e. the standard ones defined above.

Remark 1.4 The orientation of hyperkahler structures is detected by an algebraic invariant (of matrices representing the complex structures J_α), defined on generic matrices A of order $2m$ as

$$\mathcal{P}_m(A) := (1/p_m) \sum_{i_s, j_s = 1}^{2m} \varepsilon_{i_1 j_1 \ldots i_m j_m} A_{i_1 j_1} \ldots A_{i_m j_m}, \tag{1.27}$$

with $p_m = 2^m(m!)$ a combinatorial coefficient [68]. It is immediate to check that $\mathcal{P}_2(Y_\alpha) = 1$, $\mathcal{P}_2(Y_\alpha) = -1$. ⊙

Remark 1.5 As mentioned in Remark 1.3, a map $L \in SO(4)$ acting in $\mathbf{R}^4$ will induce an action on tensors which transforms the standard (positively or negatively oriented) hyperkahler structure into an equivalent one. Note that $L \in O(4)\backslash SO(4)$ (that is, in the equivalence class $\{-1\}$ for $O(4)/SO(4)$) would interchange the equivalence classes of positively and negatively oriented standard structures. $\odot$

Similarly to what we did in $\mathbf{R}^4$, we can define standard structures in $\mathbf{R}^{4n} = \mathbf{R}^4 \oplus \ldots \oplus \mathbf{R}^4$ as those which decompose as the sum of structures on each of the $\mathbf{R}^4$ component, and which correspond to a standard structure on each component.

We should however pay attention here to the fact that we could have standard structures of different orientation in each component. So while in $\mathbf{R}^4$ we have two standard structures, in $\mathbf{R}^{4n}$ we have 2^n standard structures.

Remark 1.6 As well known, we can always reduce any Riemannian metric g to the Euclidean form at a given point. Similarly, any hyperkahler structure, on any hyper-kahler manifold, can always be reduced to the standard form (of suitable orientation) *at a given point*; but this property does not extend in general to a neighborhood—no matter how small—of the point. This is readily understood recalling that the hyper-complex structures J_α are covariantly constant under the Levi-Civita connection. Thus, unless this is actually flat, they cannot be taken to be constant. Needless to say, in the Euclidean case the Levi-Civita connection is globally flat, and the reduction of the hyperkahler structures to standard form can be achieved globally. $\odot$

1.4.4 Standard Structures and Self-duality

Standard structure of positive or negative orientation in $\mathbf{R}^4$ are related to, respectively, self-dual and anti-self-dual forms. Here we will briefly discuss this point, which will be taken over again in Sect. 3.5 in greater generality.

We recall that the Hodge star operator, defined on the differential forms $\Lambda(M)$ on a manifold M (of dimension m) acts by

$$\star \; : \; \Lambda^k(M) \rightarrow \Lambda^{m-k}(M) \, .$$

Here we are interested in the case $M = \mathbf{R}^4$, and observe that for any $\beta \in \Lambda^k(\mathbf{R}^4)$, we have

$$\star \, (\star\beta) \; = \; (-1)^{k(4-k)} \, \beta \, .$$

In particular, for $k = 2$ we get $\star : \Lambda^2(\mathbf{R}^4) \rightarrow \Lambda^2(\mathbf{R}^4)$; and for $\beta \in \Lambda^2(\mathbf{R}^4)$ we just have $\star(\star\beta) = \beta$.

Thus there are $\beta \in \Lambda^2(\mathbf{R}^4)$ such that $\star\beta = \pm\beta$; these forms β are accordingly called *self-dual* or *anti-self-dual* and they span $\Lambda^2(\mathbf{R}^4)$. In other words, as well known,

$$\Lambda^2(\mathbf{R}^4) \; = \; \Lambda^2_+(\mathbf{R}^4) \oplus \Lambda^2_-(\mathbf{R}^4) \, ,$$

where Λ_+^2 (Λ_-^2) is the space of self-dual (of anti-self-dual) forms.

If now we look at the symplectic forms ω_α and $\widehat{\omega}_\alpha$ associated to the standard hyperkahler structures of, respectively, positive and negative orientation, it is immediate to check that the ω_α, and actually all the forms in the Kahler sphere **S** they generate, are *self-dual*; while the $\widehat{\omega}_\alpha$, and actually all the forms in the Kahler sphere $\widehat{\mathbf{S}}$ they generate, are *anti-self-dual*.

Similar considerations hold for the standard structures in $\mathbf{R}^{4n}$, with a relevant difference: now one should consider the restriction of the associated symplectic forms to the different $\mathbf{R}^4$ component subspaces, and on these the forms will be self-dual or anti-self-dual.

1.5 Dual Hyperkahler Structures

As well known [106], given an orientable manifold (with volume form Ω) one can consider orientation-reversing maps on it; we are in particular interested in those which reverse orientation but preserve the volume, i.e. orthogonal orientation-reversing maps. These are the maps $\varrho : M \rightarrow M$ such that $\varrho^*(\Omega) = -\Omega$.

In the case of Euclidean real spaces $(M, g) = (\mathbf{R}^m, \delta)$, the linear orthogonal orientation-reversing maps are just $O(m)/SO(m)$ (i.e. those in the equivalence class $\{-1\}$ of $O(m)/SO(m)$), as in Remark 1.5 above.

Definition 1.4 Given a hyperkahler structure $\mathbf{J} = (J_1, J_2, J_3)$ with associated hypersymplectic structure $\omega = \{\omega_1, \omega_2, \omega_3\}$ and an orthogonal orientation-reversing map ϱ, we say that the hypersymplectic structure

$$\widehat{\omega} = \{\widehat{\omega}_1, \widehat{\omega}_2, \widehat{\omega}_3\} = \{\rho^*(\omega_1), \rho^*(\omega_2), \rho^*(\omega_3)\} = \rho^*(\omega) \qquad (1.28)$$

is *dual* to ω. In view of this, we also say that ϱ is a *duality map*.

The hypersymplectic structure $\widehat{\omega}$ is associated via the Kahler relation to a hyperkahler structure $\widehat{\mathbf{J}} = (\widehat{J}_1, \widehat{J}_2, \widehat{J}_3)$; we say that this is *dual* to $\mathbf{J}$.

Note that for a given hyperkahler (hypersymplectic) structure, there are many dual ones, as there are many orthogonal orientation-reversing maps. On the other hand, all of these are in the same equivalence class. Thus for a given **Q**-structure **Q** as in (1.16), the *dual* **Q**-structure

$$\widehat{\mathbf{Q}} = \left\{ \sum_\alpha c_\alpha \widehat{J}_\alpha, \ c_\alpha \in \mathbf{R} \right\} \qquad (1.29)$$

is uniquely defined.

It is immediate to check that the **Q**-structures defined by the standard hyperkahler structures in $\mathbf{R}^4$ (of positive and negative orientation) considered above, see Sect. 1.4.3, are dual to each other.

When we consider *decomposable* hyperkahler manifolds, i.e. hyperkahler manifolds M which are the product of hyperkahler submanifolds $M_{(k)}$, we could separately reverse the orientation of each hyperkahler component, thus producing several dual hyperkahler (and hence **Q**) structures.[8] Note that these could be distinguished based on the submanifold which is orientation-reversed.[9]

If we reverse the orientation of two components $M_{(k)}$, the orientation of the full manifold M will not be changed; however we could still distinguish the two cases by the orientation of the components. Correspondingly, if we have an orthogonal map ϱ which reverse the orientation of *any* number (odd or even) of the $M_{(k)}$ components, we will consider the hypersymplectic structure $\widehat{\omega} = \varrho^*(\omega)$ as dual to ω, and analogously for the associated hyperkahler and **Q**-structures. Note that in this way we can have duality among structures with the *same* orientation.

This discussion applies in particular to the special case of hyperkahler manifolds provided by $\mathbf{R}^{4n}$ with an Euclidean structure and hyperkahler structures which are just the direct sum of standard hyperkahler structures (of either orientation) in each $\mathbf{R}^4$ subspace.

In this case a map which reverse the orientation in the k-th component will exchange the standard structures (of positive and negative orientation) in the k-th $\mathbf{R}^4$ subspace.

---

[8]In particular, for $M = \mathbf{R}^{4n} = \mathbf{R}^4 \oplus \ldots \oplus \mathbf{R}^4$, with Euclidean metric and a standard hyperkahler structure in each $\mathbf{R}^4$ component, we have a set of 2^n mutually dual hyperkahler structures.

[9]If M is hyperkahler and of dimension $m = 4n$, and it can be decomposed as the product of n hyperkahler manifolds, we say it is *fully decomposable*. This is notably the case of $\mathbf{R}^{4n}$ with standard structures, i.e. seen as the product of standard $\mathbf{R}^4$ hyperkahler manifolds.

Chapter 2
Hyperhamiltonian Dynamics

In this Chapter we introduce hyperhamiltonian dynamics and study some of the main features of hyperhamiltonian vector fields [60, 62–65, 120]. In the second part of the Chapter we will also discuss the notion of canonical transformations in hyperhamiltonian dynamics (we will see that the natural generalizations of the two equivalent ways of defining these in Hamiltonian dynamics are *not* equivalent in this context) and the relation between hyperhamiltonian vector fields and canonical transformations, generalizing a well known result in standard Hamiltonian dynamics.

2.1 The Hyperhamiltonian Evolution Equations

We start with the definition of the main character in our story, i.e. hyperhamiltonian dynamics.

Definition 2.1 Given a hyperkahler structure $\{\omega_1, \omega_2, \omega_3; g\}$ on M we define the hyperhamiltonian vector field associated to a triple of Hamiltonians $\{\mathcal{H}^1, \mathcal{H}^2, \mathcal{H}^3\}$ on M as the vector field

$$X = X_1 + X_2 + X_3 \tag{2.1}$$

where each of the X_α is the Hamiltonian vector field associated to $\mathcal{H}^\alpha$ via the symplectic form ω_α, see (1.2). That is (with no sum on α)

$$X_\alpha \lrcorner \omega_\alpha = \mathrm{d}\mathcal{H}^\alpha. \tag{2.2}$$

Remark 2.1 Note that as X is the sum of Liouville vector fields, it is also Liouville itself. ⊙

© Springer International Publishing AG 2017
G. Gaeta and M.A. Rodríguez, *Lectures on Hyperhamiltonian Dynamics and Physical Applications*, Mathematical Physics Studies, DOI 10.1007/978-3-319-54358-1_2

Let us now consider the case where local coordinates are defined in the open set $U \subseteq M$; then the symplectic forms are represented by matrices K^α,

$$\omega_\alpha = \frac{1}{2} K_{ij}^\alpha \, dx^i \wedge dx^j \,, \tag{2.3}$$

and the components f^i of the vector field

$$X = f^i(x) \, \partial_i \tag{2.4}$$

are written as

$$f^i = \sum_{\alpha=1}^{3} M_\alpha^{ij} \, (\partial_j \mathcal{H}^\alpha) \,, \tag{2.5}$$

where we have used the notation

$$M_\alpha := g^{-1} K_\alpha g^{-1} \,, \quad \text{i.e.} \quad M_\alpha^{ij} = g^{ip} K_{pq}^\alpha g^{qj} \,. \tag{2.6}$$

2.2 Hamiltonian versus Hyperhamiltonian Dynamics

A natural question immediately arises upon defining hyperhamiltonian dynamics: is this really more general than Hamiltonian one? The answer to this is positive, and can be obtained with little effort, as we show here.

First of all, we note that every Hamiltonian system is trivially hyperhamiltonian: in the hyperhamiltonian framework, it suffices to set two of the three Hamiltonian functions $\mathcal{H}^\alpha$ equal to zero to recover the standard Hamiltonian case.

On the other hand, let us check that there are systems which are hyperhamiltonian but cannot be written in Hamiltonian form with respect to *any* symplectic structure. In order to show this, we recall a result characterizing such vector fields [78].

Lemma 2.1 (Giordano-Marmo-Rubano) *Given a linear vector field*

$$X = A_j^i \, x^j \, \partial_i \,,$$

if there is $k \in N$ such that

$$\mathrm{Tr}(A^{2k+1}) \neq 0 \,,$$

then X is not Hamiltonian with respect to any symplectic structure.

The vanishing of $\mathrm{Tr}(A)$ corresponds to the condition of zero divergence, which is also satisfied by hyperhamiltonian flows. Thus in the simplest case we are looking for cases where

$$\mathcal{H}^\alpha = \frac{1}{2} \, (D_\alpha)_{ij} \, x^i x^j \tag{2.7}$$

(with D_α symmetric matrices; in the following we write all indices as lower ones to avoid confusion with powers) and

$$A := \sum_\alpha K_\alpha D_\alpha$$

satisfies $\mathrm{Tr}(A^3) = 0$.

This is obtained e.g. if [59]

$$\mathcal{H}_1 = \frac{1}{2} [x_1^2 - x_2^2 + x_3^2 - x_4^2 + 2(x_1 x_4 - x_2 x_3)] ,$$

$$\mathcal{H}_2 = \frac{1}{2}(x_1^2 + x_2^2 + x_3^2 + x_4^2) ,$$

$$\mathcal{H}_3 = 0 .$$

Thus we have shown (by explicit example) that:

Lemma 2.2 *There are hyperhamiltonian vector fields which are not Hamiltonian with respect to any symplectic structure.*

2.3 Alternative Approach to the Evolution Equations

Our definition of the hyperhamiltonian vector field was as the sum of three Hamiltonian vector fields. This is in a way not satisfactory, as such a sum does not appear to have any intrinsic meaning. It is thus appropriate to devote some page (this section) to an alternative formulation of the hyperhamiltonian evolution equations which is free from this drawback.[1]

To each symplectic form ω we associate a $(4n - 2)$-form ζ via

$$\zeta := \omega \wedge ... \wedge \omega \quad (2n - 1 \text{ factors}) . \tag{2.8}$$

In particular, to each of the three symplectic forms ω_α is thus associated a form ζ_α.

Moreover, we consider the volume form Ω in M; recall this can also be expressed, for each ω_α, as

$$\Omega = \frac{s}{(2n)!} \omega_\alpha \wedge ... \wedge \omega_\alpha \quad (s = \pm 1) . \tag{2.9}$$

---

[1] Actually, this was the *original* definition of hyperhamiltonian vector fields in [60]; our discussion in this section follows the one provided in that paper.

Then, given a hypersymplectic structure, i.e. an ordered triple of symplectic forms ω_α, to any triple of Hamiltonians $\mathcal{H}^\alpha$ is uniquely associated a vector field X defined by

$$X \lrcorner \Omega = \frac{1}{(2n-1)!} \sum_{\alpha=1}^{3} d\mathcal{H}^\alpha \wedge \zeta_\alpha. \qquad (2.10)$$

Using (2.9), this can also be rewritten as

$$X \lrcorner \sum_{\alpha=1}^{3} \omega_\alpha \wedge \zeta_\alpha = (6n\,s) \sum_{\alpha=1}^{3} d\mathcal{H}^\alpha \wedge \zeta_\alpha. \qquad (2.11)$$

Note that (2.10) defines X uniquely. On the other hand, it is immediately checked that the X defined by (2.1) and (2.2) satisfies (2.10). Thus the two definitions of X are equivalent.

Note also that (2.10) shows at once that X is Liouville, as already remarked right after Definition 2.1.

In the case where the symplectic forms admit a symplectic potential, we can consider still another form associated to any ω; recalling the definition of ζ considered above, and assuming σ is a symplectic potential for ω, we define the $(4n-1)$-form

$$\varphi := \sigma \wedge \zeta. \qquad (2.12)$$

In particular, to each of the ω_α is thus associated a $(4n-1)$-form φ_α; we will then define new forms φ and ϑ, the latter also involving the time variable t:

$$\varphi = \sum_{\alpha=1}^{3} \sigma_\alpha \wedge \zeta_\alpha \,; \quad \vartheta = \varphi + (6ns) \sum_{\alpha=1}^{3} \mathcal{H}^\alpha \left(\zeta_\alpha \wedge dt\right). \qquad (2.13)$$

We stress that $d\varphi$ is proportional to the volume form Ω, and that $d\vartheta$ is non singular. We also stress that this construction is always possible locally; and globally if $H^2(M) = 0$. Even when this condition is not satisfied, it will be possible if the symplectic forms we are considering are exact.

The forms defined in (2.13) allow to provide yet another definition of the hyperhamiltonian vector field, now seen as (the spatial component of) a vector field Z in $\tilde{M} = M \times \mathbf{R}$, where the $\mathbf{R}$ factor corresponds to the time variable. In fact, one uniquely identifies Z by

$$Z \lrcorner d\vartheta = 0, \quad Z \lrcorner dt = 1. \qquad (2.14)$$

The second condition just means that we can write

$$Z = \partial_t + Y,$$

where Y is a (possibly time-dependent) vector field on M. With this, the first equation in (2.14) decomposes (considering terms which contain or do not contain a $\mathrm{d}t$ factor) into two equations, i.e.

$$Y \lrcorner \sum_\alpha \omega_\alpha \wedge \zeta_\alpha \;=\; (6ns) \sum_\alpha \mathrm{d}\mathcal{H}^\alpha \wedge \zeta_\alpha \;;$$
$$Y \lrcorner \sum_\alpha \mathrm{d}\mathcal{H}^\alpha \wedge \zeta_\alpha \;=\; 0 . \qquad (2.15)$$

The first of these corresponds to (2.11) and, in view of the uniqueness of X, shows that in fact $Y = X$. The second equation in (2.15) is just a consequence of the first one and thus carries no further information.

Remark 2.2 The hyperhamiltonian dynamics can also be characterized in terms of a variational principle [62–64]; this is briefly discussed in Appendix B. It turns out the hyperhamiltonian dynamics defined here is also natural from the point of view of generalizing (to the quaternionic setting) the description of Hamiltonian dynamics in terms of complex analysis; this is discussed in [120].   ⊙

2.4 Hyperhamiltonian Flows and Dual Structures

In Sect. 1.5 we have introduced, for a given hyperkahler structure, the notion of dual hyperkahler structures. Here we consider the hyperhamiltonian dynamics associated to such dual structures, and (some of) its relations with the dynamics associated to the given hyperkahler structure.

2.4.1 Dual Hyperkahler Structures and Dual Hyperhamiltonian Dynamics

As discussed in Sect. 1.5, to any hyperkahler structure are associated one or more *dual* ones, with opposite orientation in M or at least in some of the minimal hyperkahler components $M_{(k)}$ of the ambient hyperkahler manifold M.

If the hyperkahler structure is characterized by the metric g and the symplectic forms $\{\omega_1, \omega_2, \omega_3\}$, any dual hyperkahler structure has the same metric and its symplectic forms will be denoted by $\{\widehat{\omega}_1, \widehat{\omega}_2, \widehat{\omega}_3\}$. The corresponding dual hyperkahler structure is then determined by g and the $\widehat{\omega}_\alpha$ through the Kahler relation.

If we have a given triple of symplectic forms $\{\omega_1, \omega_2, \omega_3\}$ and a triple of Hamiltonians $\{\mathcal{H}^1, \mathcal{H}^2, \mathcal{H}^3\}$, thus defining a hyperhamiltonian dynamical vector field

$$X \;=\; \sum X_a , \qquad X_a \lrcorner \omega_a = \mathrm{d}\mathcal{H}^a , \qquad (2.16)$$

we can consider a dual hyperhamiltonian vector field

$$\widehat{X} \;=\; \sum \widehat{X}_a \,, \quad \widehat{X}_a \,\lrcorner\, \widehat{\omega}_a = \mathrm{d}\mathcal{H}^a; \tag{2.17}$$

note this is characterized by the *same* Hamiltonians $\mathcal{H}^\alpha$ and by the dual symplectic forms $\widehat{\omega}_\alpha$.

We stress once again that there are many hyperkahler structures dual to any given one, as there are many duality maps ϱ. Thus the dual vector field to a given hyperhamiltonian vector field is surely *not* uniquely defined.

Lemma 2.3 *For X and $\widehat{X}$ as above, their commutator $Z = [X, \widehat{X}] = f^i \partial_i$ satisfies*

$$f \;=\; \big(\widehat{M}_\beta H^\beta M_\alpha \;-\; M_\beta H^\beta \widehat{M}_\alpha\big)\, \nabla\mathcal{H}^\alpha \;:=\; A_\alpha\, \nabla\mathcal{H}^\alpha \,. \tag{2.18}$$

Proof This follows by direct computation. Writing $\partial_{ij}\mathcal{H}^\alpha = H^\alpha_{ij}$ for ease of notation, we have

$$f^i = \big(M^{jk}_\alpha (\partial_k \mathcal{H}^\alpha)\big)\, \partial_j\, \big(\widehat{M}^{i\ell}_\beta (\partial_\ell \mathcal{H}^b)\big) \;-\; \big(\widehat{M}^{jk}_\alpha (\partial_k \mathcal{H}^\alpha)\big)\, \partial_j\, \big(M^{i\ell}_\beta (\partial_\ell \mathcal{H}^b)\big)$$

$$= \big(\widehat{M}^{i\ell}_\beta\, H^\beta_{\ell j}\, M^{jk}_\alpha \;-\; M^{i\ell}_\beta\, H^\beta_{\ell j}\, \widehat{M}^{jk}_\alpha\big)\, \partial_k \mathcal{H}^\alpha \,;$$

this is just our statement.   △

Remark 2.3 Note that A_α is in general not antisymmetric: in fact (using $M^T = -M$, $\widehat{M}^T = -\widehat{M}$ and $H^T = H$)

$$A^T_\alpha \;=\; -\,(\widehat{M}_\alpha H^\beta M_\beta \;-\; M_\alpha H^\beta \widehat{M}_\beta)\,,$$

and in general $A^T \neq -A$. This implies, in particular, that in general A_α is *not* a combination of the M_α and/or the $\widehat{M}_\alpha$.

On the other hand, if H^α_{ij} is a multiple of the identity—as in particular in the case of *quaternionic oscillators*, to be considered in Chap. 5—then by $[M_\alpha, \widehat{M}_\beta] = 0$ we get $[X, \widehat{X}] = 0$.   ⊙

2.4.2 Dirac Vector Fields

We have introduced the notion of dual hyperkahler structures, and correspondingly we have a notion of vector fields which are hyperhamiltonian with respect to dual hyperkahler structures.

As stressed above, dual structures are originated from maps preserving the volume form up to a sign, and reversing the orientation in any number of the minimal hyperkahler submanifolds $M_{(k)}$ in which the the hyperkahler manifold M under study can be decomposed.

Definition 2.2 Given a hyperkahler structure **J**, we say that any vector field which is hyperhamiltonian w.r.t. either **J** or any of the dual hyperkahler structures is a *Dirac vector field* for **J**.

In the same way as we usually say "hyperhamiltonian vector field" without specifying "w.r.t. the hyperkahler structure **J**", we will just speak of "Dirac vector fields".

This notion will play a substantial role in our forthcoming discussion. In particular, while in the standard Hamiltonian case the vector fields which preserve the symplectic structure are all (and only) those which are Hamiltonian—with any Hamiltonian function—under the given symplectic structure, we anticipate (see Sect. 3.2.4 for details) that in the hyperhamiltonian case the vector fields which preserve the hyperkahler form are not only those which are hyperhamiltonian w.r.t. the given hyperkahler structure, but also those which are hyperhamiltonian w.r.t. dual hyperkahler structures; that is, Dirac vector fields.

Definition 2.3 Given a hyperkahler structure **J** in (M, g), let $\widehat{\mathbf{J}}$ be the dual hyperkahler structure corresponding to a duality map which reverses orientation in *each* of the irreducible hyperkahler components of $(M, g; \mathbf{J})$. We say that a vector field which is the sum of vector fields which are hyperhamiltonian w.r.t. the **J** and the $\widehat{\mathbf{J}}$ structures is a *strictly Dirac vector field*.

2.5 First Integrals, Conservation Laws, and Poisson-like Brackets

As always when dealing with a dynamics, we are specially interested in conservation laws and first integrals for the hyperhamiltonian vector field X.

2.5.1 First Integrals

Definition 2.4 A *first integral* for the vector field X is a smooth function $F : M \to \mathbf{R}$, such that

$$\mathcal{L}_X(F) = 0 . \tag{2.19}$$

Lemma 2.4 *The smooth function $F : M \to \mathbf{R}$ is a first integral for the hyperhamiltonian vector field X if and only if the sum of its Poisson brackets with the three Hamiltonians (each w.r.t. the corresponding symplectic structure) vanishes.*

Proof The time evolution of a scalar smooth function $F : M \to \mathbf{R}$ under the hyperhamiltonian vector field X is, in view of (2.1), given by

$$\mathcal{L}_X(F) = X(F) = \sum_\alpha X_\alpha(F) \, ; \tag{2.20}$$

recalling also (2.2) and working in local coordinates, we get

$$
\begin{aligned}
\mathcal{L}_X(F) &= \sum_\alpha M_\alpha^{ij} \, (\partial_j \mathcal{H}^\alpha) \, (\partial_i F) \\
&= \sum_\alpha [(\partial_i F) \, M_\alpha^{ij} \, (\partial_j \mathcal{H}^\alpha)] = \sum_\alpha \{F, \mathcal{H}^\alpha\}_\alpha \, ,
\end{aligned}
$$

where we have denoted by $\{., .\}_\alpha$ the Poisson bracket defined by the symplectic form ω_α.   △

Remark 2.4 A special case is of course the one in which these three Poisson brackets are separately zero, but in general this is not required.   ⊙

In Hamiltonian dynamics, given two first integrals F_1 and F_2, we obtain a new first integral (which might happen to be trivial or dependent on the first two) by applying the Poisson bracket, i.e. as

$$F_3 = \{F_1, F_2\} \, .$$

It is natural to wonder if there is a result of this kind also in the framework of hyperhamiltonian dynamics; it appears this is not the case, at least if we require to have the new first integral as a homogeneous bilinear function, independent of the chosen Hamiltonians, of the derivatives of the known ones.

In fact, any such function can be written as

$$F_3 = (\partial_i F_1) \, P^{ij} \, (\partial_j F_2) \, , \tag{2.21}$$

with P a matrix.

With integration by parts, writing $X = f^i \partial_i$, and using the assumption $X(F_1) = X(F_2) = 0$, we have

$$
\begin{aligned}
X(F_3) &= [X(\partial_i F_1)] \, P^{ij} \, (\partial_j F_2) + (\partial_i F_1) \, P^{ij} \, [X(\partial_j F_2)] \\
&= [\partial_i X(F_1)] \, P^{ij} \, (\partial_j F_2) + (\partial_i F_1) \, P^{ij} \, [\partial_j X(F_2)] \\
&\quad - [(\partial_i f^k)(\partial_k F_1) \, P^{ij} \, (\partial_j F_2) + (\partial_i F_1) \, P^{ij} \, (\partial_j f^k)(\partial_k F_2)] \\
&= - [(\partial_i f^k)(\partial_k F_1) \, P^{ij} \, (\partial_j F_2) + (\partial_i F_1) \, P^{ij} \, (\partial_j f^k)(\partial_k F_2)] \, ;
\end{aligned}
$$

note we have already used the condition that F_1, F_2 are first integrals for X.

If now we write

$$\mathcal{F}^i_{\ k} = \partial_k f^i \, , \tag{2.22}$$

the above reads

$$X(F_3) = - \left[(\partial_k F_1) \, \mathcal{F}^k_{\ i} \, P^{ij} \, (\partial_j F_2) + (\partial_i F_1) \, P^{ij} \, (\mathcal{F}^T)^{\ k}_j \, (\partial_k F_2) \right]$$
$$= - \left[(\nabla F_1 \cdot \mathcal{F} P \nabla F_2) + (\nabla F_1 \cdot P \mathcal{F}^T \nabla F_2) \right]$$
$$= - \left(\nabla F_1 \cdot (\mathcal{F} P + P \mathcal{F}^T) \nabla F_2 \right) . \tag{2.23}$$

Remark 2.5 One could be tempted to require

$$\mathcal{F} P + P \mathcal{F}^T = 0 , \tag{2.24}$$

but this would actually be too much, as discussed in a moment. Moreover, an explicit computation made with the standard hyperkahler structure in $\mathbf{R}^4$ shows that this condition is satisfied only for $P = 0$.

In fact, with the standard hyperkahler structure in $\mathbf{R}^4$ and P a skew-symmetric matrix, (2.24) is a homogeneous linear system with six unknowns (the independent components of P) and six equations (and $\mathcal{F} P + P \mathcal{F}^T$ will in general be nonzero); then the only solution is $P = 0$.

Actually, the coefficient matrix (which is a 6×6 matrix) has a peculiar structure, being the sum of a symmetric plus a skew-symmetric matrix, the two having non-null elements in different places. Due to this structure, its determinant is a perfect square, in fact the square of the sum of products of second derivatives of $\mathcal{H}_i$, each product having three factors; thus it is always non-zero.   ⊙

When requiring

$$\left(\nabla F_1 \cdot (\mathcal{F} P + P \mathcal{F}^T) \nabla F_2 \right) = 0 ,$$

see (2.23) above, we should recall that F_1, F_2 are first integrals for X, i.e. their gradients are necessarily orthogonal to the stream lines of X; on the other hand, these are determined by the f^i and are hence embodied in the matrix $\mathcal{F}$.

Thus we should actually require that

$$\left(\xi \cdot (\mathcal{F} P + P \mathcal{F}^T) \eta \right) = 0 \tag{2.25}$$

for all vectors ξ, η orthogonal to the kernel of X. This requirement is weaker than (2.24).

2.5.2 Conservation Laws

By a *conservation law* we mean a "conserved form of submaximal degree" (this corresponds via Hodge duality to a conserved vector, similar to the Runge-Lenz vector for the Kepler problem), i.e. an object with $m = 4n$ components which is preserved under X. Rather than seeing this as a vector on M, it is more convenient to adopt the dual point of view and see it as a differential form. This in turn can be a one-form $\theta \in \Lambda^1(M)$ or, using Hodge duality [123, 124], a form of submaximal degree, $\Theta \in \Lambda^{(4n-1)}(M)$. It turns out that the latter point of view is often somehow more convenient.

Definition 2.5 A *conservation law* for the vector field X in the $4n$-dimensional manifold M is a form $\Theta \in \Lambda^{(4n-1)}(M)$ such that $\mathcal{L}_X(\Theta) = 0$.

In particular, given a triple of symplectic structures ω_α, to any triple $\mathcal{H}^\alpha$ of Hamiltonians (which uniquely define a hyperhamiltonian vector field X) is canonically associated a conserved form $\Theta \in \Lambda^{(4n-1)}(M)$; this is just

$$\Theta := \sum_\alpha d\mathcal{H}^\alpha \wedge \zeta_\alpha . \tag{2.26}$$

Lemma 2.5 *The form Θ is preserved by the vector field X defined by the Hamiltonians $\mathcal{H}^\alpha$,*

$$\mathcal{L}_X(\Theta) = 0 . \tag{2.27}$$

Proof In fact, Θ is obviously closed; thus

$$\mathcal{L}_X(\Theta) = d(X \lrcorner \Theta).$$

The form of Θ and the equations (2.1), (2.2) yield

$$(X \lrcorner \Theta) = (2n - 1)! \, [X \lrcorner (X \lrcorner \Omega)] = 0 ;$$

thus *a fortiori* $d(X \lrcorner \Theta) = 0$ and hence (2.27).   △

Remark 2.6 Actually, the construction via Hodge duality canonically associates a $(4n - 1)$-form χ to any vector field Y on M; this is written as

$$Y \lrcorner \Omega = \chi. \tag{2.28}$$

Conversely, this relation associates a vector field Y to any form $\chi \in \Lambda^{(4n-1)}(M)$; that is, there is a map $F : \Lambda^{(4n-1)}(M) \to \mathcal{X}(M)$.

Given two forms $\alpha, \beta \in \Lambda^{(4n-1)}(M)$, we have vector fields $Y_\alpha = F(\alpha)$ and $Y_\beta = F(\beta)$ on M. We can then take the commutator of these vector fields, and consider the form in $\Lambda^{(4n-1)}(M)$ associated to it, i.e. the form

$$\gamma = F^{-1}([F(\alpha), F(\beta)]) . \tag{2.29}$$

This construction defines a natural (antisymmetric) binary operation $\{.,.\}$ on $\Lambda^{(4n-1)}(M)$, so that the above can be rewritten as

$$\gamma = \{\alpha, \beta\} . \tag{2.30}$$

This is just the corresponding of the commutator when considering (both) the duality between vector fields and one forms, and the Hodge duality between $\Lambda^1(M)$ and $\Lambda^{(4n-1)}(M)$.   ⊙

Remark 2.7 The construction of the above remark can be used to generate new conservation laws from known ones. That is, if α and β are conserved $(4n-1)$-forms, then $\gamma = \{\alpha, \beta\}$ is also conserved. In this respect, as mentioned in [60] (see Sect. 3 in there), $\{., .\}$ is reminiscent of the Poisson bracket.

It should be stressed, however, that $\{., .\}$ is just based on standard and Hodge duality; hence it only uses the metric structure in M and its volume form, and *not* the symplectic or hyperkahler structures.

In fact, a vector field Y_i is preserved under X if $\mathcal{L}_X(Y) = 0$; but

$$\mathcal{L}_X(Y) = [X, Y] \, ;$$

thus the conservation of γ just amounts to the fact that $[X, Y_\alpha] = 0$ and $[X, Y_\beta] = 0$ entail $[X, [Y_\alpha, Y_\beta]] = 0$. In other words the fact that, for γ as in (2.30), $\mathcal{L}_X(\alpha) = 0 = \mathcal{L}_X(\beta)$ entails $\mathcal{L}_X(\gamma) = 0$ is just a consequence of the Jacobi identity.   ⊙

2.5.3 Combining First Integrals and Conservation Laws

If we have a first integral F and a conservation law $\Xi \in \Lambda^{(4n-1)}(M)$, we can readily produce a new scalar conserved quantity, i.e. a new first integral (as in the Hamiltonian case, this might be dependent on the quantities mentioned above, or even turn out to be trivial).

Lemma 2.6 *Let X be a hyperhamiltonian vector field in M, Ω the volume form in M, $\Xi \in \Lambda^{(4n-1)}(M)$ a conservation law for X, F a conserved quantity for X, and $\phi = \mathrm{d}F$. Then the scalar quantity σ defined by*

$$\Xi \wedge \phi = \sigma \Omega \tag{2.31}$$

is a first integral for X.

Proof In fact, $X(F) = 0$ can also be written as $X \lrcorner \mathrm{d}F = 0$; thus $\phi := \mathrm{d}F \in \Lambda^1(M)$ is a conserved one-form. As the form $\Xi \wedge \phi \in \Lambda^{(4n)}(M)$ is of maximal degree, this defines indeed a scalar function $\sigma : M \to \mathbf{R}$ through (2.31).

Taking into account that X is Liouville, we readily have

$$\mathcal{L}_X(\Xi \wedge \phi) = X(\sigma) \Omega + \sigma \mathcal{L}_X(\Omega) = X(\sigma) \Omega \, ; \tag{2.32}$$

thus the form $\Xi \wedge \phi$ is conserved if and only if $X(\sigma) = 0$, i.e. if σ is a first integral for X. On the other hand, by assumption Ξ and ϕ are both conserved forms, hence

$$\mathcal{L}_X(\Xi \wedge \phi) \;=\; [\mathcal{L}_X(\Xi)] \wedge \phi \;-\; \Xi \wedge [\mathcal{L}_X(\phi)] \;=\; 0 \;. \tag{2.33}$$

That is, σ defined by (2.31) is indeed a first integral for X. $\triangle$

2.6   The Hyperkahler Form

2.6.1   *General Setting*

We have seen above that any hyperhamiltonian vector field is Liouville, i.e. it preserves the volume form Ω on M, see Remark 2.1.

On the other hand, each of the X_α, see (2.1) and (2.2), preserves the corresponding symplectic form[2] ω_α,

$$\mathcal{L}_{X_\alpha}(\omega_\alpha) \;=\; 0 \quad (\alpha = 1, 2, 3) \;, \tag{2.34}$$

but it will in general *not* preserve the other two symplectic forms, $\mathcal{L}_{X_\alpha}(\omega_\beta) \neq 0$ for $\alpha \neq \beta$. This also means that—except in very special cases, i.e. for very special choices of the Hamiltonians $\mathcal{H}^\alpha$—the hyperhamiltonian vector field X will not preserve the three symplectic forms: the hyperhamiltonian dynamics is in general *not* three-holomorphic.

One can and should also consider the *hyperkahler four-form*

$$\Psi \;=\; \frac{1}{2} \sum_{\alpha=1}^{3} \omega_\alpha \wedge \omega_\alpha \;. \tag{2.35}$$

Lemma 2.7 *Equivalent hyperkahler structures have the same hyperkahler four-form.*

Proof Let us consider a set of different forms $\widetilde{\omega}_\alpha$ obtained from the set ω_α by an SO(3) rotation:

$$\widetilde{\omega}_\alpha \;=\; \sum_{\beta=1}^{3} R_{\alpha\beta} \, \omega_\beta \;. \tag{2.36}$$

We can compute the hyperkahler four-form, which we will denote by $\widetilde{\Psi}$, based on these symplectic forms, and show it is just the same as the one based on the ω_α forms. In fact,

---

[2]And hence is canonical (in the sense of standard Hamiltonian dynamics) for the corresponding symplectic form.

$$\widetilde{\Psi} = \frac{1}{2} \sum_{\alpha=1}^{3} \widetilde{\omega}_\alpha \wedge \widetilde{\omega}_\alpha = \frac{1}{2} \sum_{\alpha=1}^{3} \left(\sum_{\beta=1}^{3} R_{\alpha\beta} \omega_\beta \right) \wedge \left(\sum_{\gamma=1}^{3} R_{\alpha\gamma} \omega_\gamma \right)$$

$$= \frac{1}{2} \sum_{\alpha,\beta,\gamma=1}^{3} R_{\alpha\beta} R_{\alpha\gamma} \omega_\beta \wedge \omega_\gamma = \frac{1}{2} \sum_{\beta,\gamma=1}^{3} \delta_{\beta\gamma} \omega_\beta \wedge \omega_\gamma$$

$$= \frac{1}{2} \sum_{\beta=1}^{3} \omega_\beta \wedge \omega_\beta = \Psi$$

This concludes the proof. $\triangle$

Remark 2.8 We have thus shown that Ψ is actually not associated to the given triple of symplectic form but to the **Q**-structure these identify. $\odot$

In the case of fully decomposable hyperkahler manifolds one can show that if two hyperkahler structures (induce the same decomposition of the manifold and) have the same hyperkahler form, then are equivalent. This will be done in the following subsection.

2.6.2 Hyperkahler Form for Low-Dimensional Standard Structures

It may be worth giving explicit expression for the standard hyperkahler structures in $\mathbf{R}^4$ and in $\mathbf{R}^8$, see Sect. 1.4.3 for the explicit expression of the ω_α and $\widehat{\omega}_\alpha$ in this case.

By a trivial computation, we obtain that the hyperkahler forms $\Psi^{(\pm)}$ for the positively and negatively oriented standard hyperkahler structure in $\mathbf{R}^4$ are

$$\Psi^{(\pm)} = \pm 3 \left(dx^1 \wedge dx^2 \wedge dx^3 \wedge dx^4 \right) . \tag{2.37}$$

The situation in $\mathbf{R}^8$ is slightly more complex, but also more representative of the general situation in $\mathbf{R}^{4n}$ (this is why we report the explicit formulas below). We will denote the basis symplectic structures as $\omega_\alpha^{(\pm\pm)}$, where the upper indices refer to the orientation in the two basic four-dimensional blocks. Thus we will have in explicit terms

$$\omega_1^{(++)} = dx^1 \wedge dx^2 + dx^3 \wedge dx^4 + dx^5 \wedge dx^6 + dx^7 \wedge dx^8 ,$$
$$\omega_2^{(++)} = dx^1 \wedge dx^4 + dx^2 \wedge dx^3 + dx^5 \wedge dx^8 + dx^6 \wedge dx^7 ,$$
$$\omega_3^{(++)} = dx^1 \wedge dx^3 + dx^2 \wedge dx^4 + dx^5 \wedge dx^7 - dx^6 \wedge dx^8 ;$$
$$\omega_1^{(+-)} = dx^1 \wedge dx^2 + dx^3 \wedge dx^4 + dx^5 \wedge dx^6 - dx^7 \wedge dx^8 ,$$
$$\omega_2^{(+-)} = dx^1 \wedge dx^4 + dx^2 \wedge dx^3 + dx^5 \wedge dx^8 - dx^6 \wedge dx^7 ,$$
$$\omega_3^{(+-)} = dx^1 \wedge dx^3 + dx^2 \wedge dx^4 + dx^5 \wedge dx^7 + dx^6 \wedge dx^8 ;$$

$$\omega_1^{(-+)} = dx^1 \wedge dx^2 - dx^3 \wedge dx^4 + dx^5 \wedge dx^6 + dx^7 \wedge dx^8 \,,$$
$$\omega_2^{(-+)} = dx^1 \wedge dx^4 - dx^2 \wedge dx^3 + dx^5 \wedge dx^8 + dx^6 \wedge dx^7 \,,$$
$$\omega_3^{(-+)} = dx^1 \wedge dx^3 - dx^2 \wedge dx^4 + dx^5 \wedge dx^7 - dx^6 \wedge dx^8 \,;$$
$$\omega_1^{(--)} = dx^1 \wedge dx^2 - dx^3 \wedge dx^4 + dx^5 \wedge dx^6 - dx^7 \wedge dx^8 \,,$$
$$\omega_2^{(--)} = dx^1 \wedge dx^4 - dx^2 \wedge dx^3 + dx^5 \wedge dx^8 - dx^6 \wedge dx^7 \,,$$
$$\omega_3^{(--)} = dx^1 \wedge dx^3 - dx^2 \wedge dx^4 + dx^5 \wedge dx^7 + dx^6 \wedge dx^8 \,.$$

We will correspondingly write

$$\Psi^{(s_1 s_2)} = \frac{1}{2} \sum_{\alpha=1}^{3} \omega_\alpha^{(s_1 s_2)} \wedge \omega_\alpha^{(s_1 s_2)} \,.$$

By simple (and rather boring) computations we get the explicit expressions for the $\Psi^{(s_1 s_2)}$:

$$\begin{aligned}
\Psi^{(++)} = \ & 3 \left(dx^1 \wedge dx^2 \wedge dx^3 \wedge dx^4 + dx^5 \wedge dx^6 \wedge dx^7 \wedge dx^8 \right) \\
& + dx^1 \wedge dx^2 \wedge dx^5 \wedge dx^6 + dx^1 \wedge dx^2 \wedge dx^7 \wedge dx^8 \\
& + dx^1 \wedge dx^3 \wedge dx^5 \wedge dx^7 - dx^1 \wedge dx^3 \wedge dx^6 \wedge dx^8 \\
& + dx^1 \wedge dx^4 \wedge dx^5 \wedge dx^8 + dx^1 \wedge dx^4 \wedge dx^6 \wedge dx^7 \\
& + dx^2 \wedge dx^3 \wedge dx^5 \wedge dx^8 + dx^2 \wedge dx^3 \wedge dx^6 \wedge dx^7 \\
& - dx^2 \wedge dx^4 \wedge dx^5 \wedge dx^7 + dx^2 \wedge dx^4 \wedge dx^6 \wedge dx^8 \\
& + dx^3 \wedge dx^4 \wedge dx^5 \wedge dx^6 + dx^3 \wedge dx^4 \wedge dx^7 \wedge dx^8 \,; \\
\Psi^{(+-)} = \ & 3 \left(dx^1 \wedge dx^2 \wedge dx^3 \wedge dx^4 - dx^5 \wedge dx^6 \wedge dx^7 \wedge dx^8 \right) \\
& + dx^1 \wedge dx^2 \wedge dx^5 \wedge dx^6 - dx^1 \wedge dx^2 \wedge dx^7 \wedge dx^8 \\
& + dx^1 \wedge dx^3 \wedge dx^5 \wedge dx^7 + dx^1 \wedge dx^3 \wedge dx^6 \wedge dx^8 \\
& + dx^1 \wedge dx^4 \wedge dx^5 \wedge dx^8 - dx^1 \wedge dx^4 \wedge dx^6 \wedge dx^7 \\
& + dx^2 \wedge dx^3 \wedge dx^5 \wedge dx^8 - dx^2 \wedge dx^3 \wedge dx^6 \wedge dx^7 \\
& - dx^2 \wedge dx^4 \wedge dx^5 \wedge dx^7 - dx^2 \wedge dx^4 \wedge dx^6 \wedge dx^8 \\
& + dx^3 \wedge dx^4 \wedge dx^5 \wedge dx^6 - dx^3 \wedge dx^4 \wedge dx^7 \wedge dx^8 \,; \\
\Psi^{(-+)} = \ & {-3} \left(dx^1 \wedge dx^2 \wedge dx^3 \wedge dx^4 - dx^5 \wedge dx^6 \wedge dx^7 \wedge dx^8 \right) \\
& + dx^1 \wedge dx^2 \wedge dx^5 \wedge dx^6 + dx^1 \wedge dx^2 \wedge dx^7 \wedge dx^8 \\
& + dx^1 \wedge dx^3 \wedge dx^5 \wedge dx^7 - dx^1 \wedge dx^3 \wedge dx^6 \wedge dx^8 \\
& + dx^1 \wedge dx^4 \wedge dx^5 \wedge dx^8 + dx^1 \wedge dx^4 \wedge dx^6 \wedge dx^7 \\
& - dx^2 \wedge dx^3 \wedge dx^5 \wedge dx^8 - dx^2 \wedge dx^3 \wedge dx^6 \wedge dx^7 \\
& + dx^2 \wedge dx^4 \wedge dx^5 \wedge dx^7 - dx^2 \wedge dx^4 \wedge dx^6 \wedge dx^8 \\
& - dx^3 \wedge dx^4 \wedge dx^5 \wedge dx^6 - dx^3 \wedge dx^4 \wedge dx^7 \wedge dx^8 \,;
\end{aligned}$$

$$\Psi^{(--)} = -3 \left(dx^1 \wedge dx^2 \wedge dx^3 \wedge dx^4 + dx^5 \wedge dx^6 \wedge dx^7 \wedge dx^8 \right)$$
$$+ dx^1 \wedge dx^2 \wedge dx^5 \wedge dx^6 - dx^1 \wedge dx^2 \wedge dx^7 \wedge dx^8$$
$$+ dx^1 \wedge dx^3 \wedge dx^5 \wedge dx^7 + dx^1 \wedge dx^3 \wedge dx^6 \wedge dx^8$$
$$+ dx^1 \wedge dx^4 \wedge dx^5 \wedge dx^8 - dx^1 \wedge dx^4 \wedge dx^6 \wedge dx^7$$
$$- dx^2 \wedge dx^3 \wedge dx^5 \wedge dx^8 + dx^2 \wedge dx^3 \wedge dx^6 \wedge dx^7$$
$$+ dx^2 \wedge dx^4 \wedge dx^5 \wedge dx^7 + dx^2 \wedge dx^4 \wedge dx^6 \wedge dx^8$$
$$- dx^3 \wedge dx^4 \wedge dx^5 \wedge dx^6 + dx^3 \wedge dx^4 \wedge dx^7 \wedge dx^8 .$$

The net message to be extracted from these fully explicit formulas is that (at difference with possible, but over-optimistic, expectations) already for $n = 2$, and hence *a fortiori* for general $n > 1$, there is no simple relation—that is, no relation amounting just to an overall sign switch—between the hyperkahler forms of mutually dual hyperkahler structures.

This also provide evidence (albeit not a formal proof, see below for that) for the following statement:

Lemma 2.8 *Two hyperkahler structures in Euclidean $\mathbf{R}^{4n}$, generating the same splitting of $\mathbf{R}^{4n} = \mathbf{R}^4 \oplus \dots \oplus \mathbf{R}^4$ into $\mathbf{R}^4$ invariant subspaces, have the same hyperkahler form if and only if they are equivalent.*

Proof As we have seen above, any hyperkahler structure in $(\mathbf{R}^{4n}, \delta)$ can be reduced to a standard one, i.e. to one which is the direct sum of positively or negatively oriented one in each of the $\mathbf{R}^4$ components.

The statement can be (and was, see above) explicitly checked for $n = 1$. In the case $n > 1$ the only possibility to have non-equivalent structures sharing the same hyperkahler form is through sign compensations, which was ruled out by our explicit computation for $n = 2$. In more formal terms if two hyperkahler structures $\mathbf{J}_1$ and $\mathbf{J}_2$ in $\mathbf{R}^{4n}$ (with $n \geq 2$) share the same hyperkahler form, i.e. (with obvious notation) $\Psi_1 = \Psi_2 = \Psi$, we can just consider the restriction of $\mathbf{J}_k$ and Ψ_k to a $\mathbf{R}^4$ subspace, invariant under $\mathbf{J}_k$. We have seen above that the restrictions of Ψ_k will coincide if and only if the restrictions of the $\mathbf{J}_k$ are equivalent. Thus if $\Psi_1 = \Psi_2$, and hence their restrictions also coincide for any choice of the $\mathbf{R}^4$ subspace, the $\mathbf{J}_1$ and $\mathbf{J}_2$ are equivalent in any $\mathbf{R}^4$ subspace, and hence for the full $\mathbf{R}^{4n}$ space. △

2.7 Hyperkahler and Canonical Maps

It is well known that in the standard Hamiltonian case, Hamiltonian vector fields generate a one-parameter (local) group of symplectic, i.e. canonical, transformations [10, 12, 109]; if the vector field is complete, we have a global group.

We will generalize this result (in this case, with substantial differences) to the hyperhamiltonian case.[3] It will actually turn out that there are *two* generalization of

---

[3]The discussion of this Section will follow our paper [68].

the notion of canonical transformations to the hyperhamiltonian framework; these will be called *canonical* and *hyperkahler* maps. In the next Chap. 3 we will discuss in detail canonical maps, while the study of hyperkahler ones is postponed to the subsequent Chap. 4.

We start by recalling the relevant notions and results in the standard Hamiltonian case.

2.7.1 Symplectic and Canonical Maps in Standard Hamiltonian Dynamics

Let (M, ω) be a symplectic manifold (of dimension $2n$); we say that a map $\phi : M \to M$ is *symplectic* if it preserves the symplectic form ω, i.e. if

$$\phi^*(\omega) = \omega . \tag{2.38}$$

An equivalent characterization is also quite common (we refer e.g. to Sect. 44 of [10] for detail). As well known, by Darboux theorem [10] one can introduce local coordinates (p_a, q^a) (for $a = 1, ..., n$) in a neighborhood $U \subset M$, such that $\omega = \mathrm{d}p_a \wedge \mathrm{d}q^a$. Then, one considers local manifolds of minimal dimension on which ω is non-degenerate; these are two-dimensional and are spanned by q^a and p_a (with same a). They are known as *Darboux submanifolds* and denoted as U_a; these also correspond to leaves of the Abelian distribution generated by the Hamiltonian vector fields associated with canonical coordinates.

Let us consider a given point in U and the manifolds U_a through this. Denote by ι_a the embedding $\iota_a : U_a \hookrightarrow U \subseteq M$; then the restriction $\iota_a^*\omega$ of the symplectic form to U_a provides a volume form $\Omega_a = \mathrm{d}p_a \wedge \mathrm{d}q^a$ (no sum on a) on U_a. Then, for any two-chain A in U and with $\pi_a A$ the projection of A to U_a,

$$\int_A \omega = \int_A \sum_{a=1}^n \mathrm{d}p_a \wedge \mathrm{d}q^a = \sum_{a=1}^n \int_A \Omega_a = \sum_{a=1}^n \mathrm{area}[\pi_a A] ;$$

thus preservation of ω is *equivalent* to preservation of the sum of oriented areas of projection of any A to Darboux submanifolds. That is, the map ϕ is canonical if and only if

$$\sum_{a=1}^n \mathrm{area}[\pi_a A] = \sum_{a=1}^n \mathrm{area}[\pi_a(\phi A)] .$$

It should be noted that if we start from a manifold equipped with a Riemannian metric, passing to Darboux coordinates will in general not preserve the representation of the metric tensor in coordinates, i.e. not preserve the (matrix g_{ij} representing the) metric. Thus this construction is general not viable if one requires preservation of the metric, as in the Kahler case.

In the case of a Kahler manifold, the symplectic form ω corresponds to a complex structure J through the Kahler relation (1.14). This satisfies $J^2 = -I$, and provides a splitting of $T_0 M$ (at any point $m_0 \in M$) into two-dimensional invariant subspaces; the volume form Ω defined in M induces volume forms Ω_a in each of these, and $\omega = \sum \Omega_a$. Thus again canonical transformations can be characterized as those satisfying[4]

$$\sum_{a=1}^{n} \Omega_a = \sum_{a=1}^{n} \phi^*(\Omega_a) . \tag{2.39}$$

Remark 2.9 Note this construction does *not* make use of Darboux coordinates or submanifolds, but only of the splitting of TM induced by the action of the complex structure; moreover, we only consider volume forms.   ⊙

2.7.2 Hyperkahler Maps

Let us now pass to consider hyperhamiltonian dynamics. We will first focus our attention on maps which preserve the **Q**-structure; as we have seen just above this amounts to mapping a hyperkahler structure into an equivalent one. We will thus use the name *hyperkahler* for such maps (we will also use the name **Q**-*map*).

Definition 2.6 Let $(M, g; \mathbf{J})$ be a hyperkahler manifold. We say that the orthogonal map $\phi : M \to M$ is *hyperkahler*—or a **Q**-*map*—if it maps the hyperkahler structure into an equivalent one, i.e. if $\phi^* : \mathbf{S} \to \mathbf{S}$. It is *strongly hyperkahler* if it leaves the three complex structures J_α invariant, $\phi^*(J_\alpha) = J_\alpha$ for $\alpha = 1, 2, 3$.

These also have, of course, a symplectic counterpart:

Definition 2.6' Let $(M, g; \omega_1, \omega_2, \omega_3)$ be a hypersymplectic manifold. We say that the orthogonal map $\phi : M \to M$ is *hypersymplectic* if it maps the hypersymplectic structure into an equivalent one, i.e. if $\phi^* : \mathcal{S} \to \mathcal{S}$. It is *strongly hypersymplectic* if it leaves the hypersymplectic structures invariant, i.e. if $\phi^*(\omega_\alpha) = \omega_\alpha$ for $\alpha = 1, 2, 3$.

It is easily seen that a map is hyperkahler if and only if it is hypersymplectic (recall that such maps are required to be orthogonal, i.e. to preserve g).

The two definitions above should be seen as the generalization (from the point of view of complex and symplectic structures respectively) to the hyperkahler framework of the familiar condition of preservation of the symplectic form, relevant in Symplectic Geometry and Hamiltonian Mechanics [10, 12, 30, 79, 82, 109].

Remark 2.10 In the standard Hamiltonian dynamics framework, one also considers the condition to preserve the sum of the oriented areas of projections to the Darboux submanifolds. In the hyperkahler context, the generalization of this condition presents

---

[4]The reader is referred to Arnold [10] (see in particular Sect. 41.E there) for a discussion of the interrelations between orthogonal, symplectic and unitary transformations.

an obvious problem: that is, now there is no analogue of Darboux theorem, and hence no natural notion of Darboux manifolds.   ⊙

Remark 2.11 Note that a Hamiltonian flow (i.e. a hyperhamiltonian flow in which only one of the three Hamiltonians is non-zero) will generate a one-parameter group of hyperkahler maps, but these are *not* strongly hyperkahler. To see this, it suffices to consider the case with $(M, g) = (\mathbf{R}^4, \delta)$ with standard hyperkahler structure and $\mathcal{H}_1 = |\beta|^2/2$, $\mathcal{H}_2 = \mathcal{H}_3 = 0$. Then ω_1 is preserved, while ω_2 and ω_3 change according to

$$\widetilde{\omega}_2 = \cos(\theta)\omega_2 - \sin(\theta)\omega_3, \quad \widetilde{\omega}_3 = \sin(\theta)\omega_2 + \cos(\theta)\omega_3 \ ;$$

here θ is an angle, linearly depending on time. Thus the forms ω_2, ω_3 are rotated in the plane they span in $\mathcal{Q}$. In other words, the hypersymplectic structure (and hence the hyperkahler one) is in this case mapped into an equivalent—but different—one. This shows indeed we have hyperkahler, but not strongly hyperkahler, maps.   ⊙

2.7.3 Canonical Maps

The problem mentioned in Remark 2.10 is not present for Euclidean hyperkahler manifolds. In fact, in this case the complex structures (which, as mentioned above, see Remark 1.6 in Chap. 1, can always be reduced at a single point—and globally in the Euclidean case—to the standard forms seen in Sect. 1.4.3) provide a natural splitting of $TM = \mathbf{R}^{4n}$ into the sum of $\mathbf{R}^4$ subspaces,

$$\mathbf{R}^{4n} = \mathbf{R}^4_{(1)} \oplus ... \oplus \mathbf{R}^4_{(n)} \ .$$

We will denote by ι_k the embedding of $\mathbf{R}^4_{(k)}$ into $\mathbf{R}^{4n}$; thus $\iota_k^* : \Lambda(\mathbf{R}^{4n}) \to \Lambda(\mathbf{R}^4_{(k)})$ will represent the restriction of forms in $\Lambda(\mathbf{R}^{4n})$ to the $\mathbf{R}^4_{(k)}$ subspace.

Correspondingly, given a symplectic form ω we will write

$$\omega^{(k)} := \iota_k^*(\omega) \quad (k = 1, ..., n) \ . \tag{2.40}$$

Note that $\iota_k^*(\omega \wedge \omega) = \omega^{(k)} \wedge \omega^{(k)}$ (no sum on k); this is the volume form $\Omega^{(k)}$ in the $\mathbf{R}^4_{(k)}$ subspace.

Remark 2.12 As just recalled, our construction and in particular the splitting of $T_{m_0}M$ can be done *at a single point* $m_0 \in M$ for any manifold, not just Euclidean ones. The obstacle to an extension of this construction to the general case is not only that the splitting is ill-defined in any neighborhood, no matter how small, of the reference point m_0: in fact, even at the reference point the splitting is in general not invariant under the holonomy group, and thus not intrinsic (see also the discussion in Sects. 2.7.3 and 3.2.3, as well as Appendix A). This problem is of course absent for the Euclidean case, where the holonomy group reduces to the identity.

Note also that in this respect any holonomy action in $Sp(1)$ amounts to mapping the hyperkahler structure into an equivalent one; and any action in $Sp(1) \times ... \times Sp(1) \subset Sp(n)$, where each of the $Sp(1)$ factors is acting in one of the irreducible hyperkahler components of $T_{m_0}M \simeq \mathbf{R}^{4n} = \mathbf{R}^4 \oplus ... \oplus \mathbf{R}^4$ (the splitting being of course that induced by the $\mathbf{J}$ structure, see above) does not alter the splitting of $TM = \mathbf{R}^{4n}$ into the sum of $\mathbf{R}^4$ subspaces, hence the $\Omega^{(a)}$ four-forms.

In other words, our subsequent discussion will apply not only to the case where the holonomy group reduces to the identity (as for Euclidean spaces) but will also apply to the cases where the holonomy group lies in $Sp(1) \times ... \times Sp(1)$.

This is maybe an appropriate point to also recall that, by the Ambrose-Singer theorem [5, 106, 112, 123, 135] (see Appendix A for a statement), the holonomy group is generated by the curvature of the connection; thus if the Levi-Civita connection on (M, g) has a curvature form lying in $sp(1) \times ... \times sp(1)$, we are guaranteed to be in the case where our discussion applies.   ⊙

We are now ready to give our definition of canonical maps for Euclidean hyperkahler manifolds, which can be given in two equivalent ways.

Definition 7a Let $(M = \mathbf{R}^{4n}, g = \delta; \mathbf{J})$ be an Euclidean hyperkahler manifold, with ι_k $(k = 1, ..., n)$ as above, and $\mathcal{S}$ the corresponding symplectic Kahler sphere. We say that the map $\phi : M \to M$ is *canonical* for the hyperkahler structure $(g = \delta; \mathbf{J})$ if, for any $\omega \in \mathcal{S}$ and any $k = 1, .., n$, it satisfies

$$\iota_k^*[\phi^*(\omega \wedge \omega)] = \iota_k^*(\omega \wedge \omega) \equiv \omega^{(k)} \wedge \omega^{(k)} . \tag{2.41}$$

Definition 7b Let $(M = \mathbf{R}^{4n}, g = \delta; \mathbf{J})$ be an Euclidean hyperkahler manifold, with ι_k $(k = 1, ..., n)$ as above, and let ω_α be the symplectic structures associated to the J_α. The map $\phi : M \to M$ is *canonical* for the hyperkahler structure $(g = \delta; \mathbf{J})$ if (with no sum on k), for any $\alpha = 1, 2, 3$ and $k = 1, ..., n$,

$$\iota_k^*[\phi^*(\omega_\alpha \wedge \omega_\alpha)] = \iota_k^*(\omega_\alpha \wedge \omega_\alpha) . \tag{2.42}$$

A stronger notion of canonicity, which implies the previous one, can also be defined (and was proposed in earlier works of ours [70]), but turns out to be too restrictive and hence of little use (see Appendix A in [71]); this is given below for the sake of completeness.

Definition 2.8 Let $(M, g; \mathbf{J})$ be a hyperkahler manifold, and $\mathcal{Q}$ the corresponding symplectic Kahler sphere. We say that the map $\phi : M \to M$ is *strongly canonical* if, for any $\omega \in \mathcal{S}$, it preserves the form $\omega \wedge \omega$, i.e. $\phi^*(\omega \wedge \omega) = \omega \wedge \omega$. Equivalently, if and only if (with no sum on a)

$$\phi^*(\omega_\alpha \wedge \omega_\alpha) = \omega_\alpha \wedge \omega_\alpha \quad \alpha = 1, 2, 3 .$$

As mentioned above, it is immediate to check that if a map is strongly canonical it is also canonical, while the converse is not true.

Remark 2.13 It is clear that the two notions of canonical and hyperkahler maps (or **Q**-maps) proposed here are *not* equivalent (at difference with the notion holding in the symplectic or Kahler case which they generalize). In a way, **Q**-maps preserve the **Q**-structure (that is, the hyperkahler form), while canonical ones only preserve the restriction of this to any irreducible hyperkahler component; moreover, note that we are *not* requiring canonical maps to be orthogonal.

Consider e.g. the standard ω_1 in $(\mathbf{R}^4, \delta)$ (see Sect. 1.4.3): under the map $x^1 \to \lambda x^1$, $x^2 \to \lambda x^2$, $x^3 \to \lambda^{-1} x^3$, $x^4 \to \lambda^{-1} x^4$, the form ω_1 is not preserved (note g is not preserved as well) nor mapped to a different form in $\mathcal{S}$, but $\omega_1 \wedge \omega_1$ is invariant (the forms ω_2 and ω_3 are instead invariant themselves, and *a fortiori* we get invariance of $\omega_2 \wedge \omega_2$ and of $\omega_3 \wedge \omega_3$). More generally, a canonical map could even mix the positively and negatively oriented structures.   ⊙

Remark 2.14 Our Definition implies that canonical vector fields preserve the volume forms $\Omega^{(k)} = (1/2)(\omega^{(k)} \wedge \omega^{(k)})$ (no sum on k) in the four dimensional $\mathbf{R}^4_{(k)}$ subspaces.[5]   ⊙

Remark 2.15 A hyperkahler map transform the triple of the ω_α into an equivalent one, and hence preserves the hyperkahler four-form $\Phi = (1/2)(\omega \wedge \omega)$. Obviously preservation of the form $\omega \wedge \omega$ implies preservation of its restriction to any subspace, and we immediately have that hyperkahler maps are also canonical. The converse is not true, as shown by the simple explicit example in Remark 2.13 above. Note also that the definition of canonical transformation does *not* require to preserve the metric g; canonical maps which are not orthogonal are definitely not **Q**-maps (the map considered in Remark 2.13 is indeed an example of this case).   ⊙

The relation between hyperhamiltonian dynamics and hyperkahler or canonical transformations will be discussed in detail in the following two Chapters.

We anticipate that we have a *partial* generalization of the familiar results holding in the Hamiltonian case, where any Hamiltonian vector field generates a one-parameter (local) group of canonical transformations, and any such group admits a Hamiltonian vector field as generator.

In fact, in the hyperhamiltonian case we will obtain that any hyperhamiltonian vector field generates a one-parameter (local) group of canonical transformations, and that any such group admits a Dirac vector field (rather than a hyperhamiltonian one) as generator. Moreover—as already stressed—in the hyperkahler case hyperkahler and canonical transformations (in the sense of the definitions above) are *not* the same.

Remark 2.17 Finally, we note that $\Psi \wedge \dots \wedge \Psi$ (with n factors) is proportional to the volume form Ω; thus preservation of Ψ implies, once again, preservation of Ω (the converse is in general not true).

We will also see that in Euclidean $\mathbf{R}^4$ with standard hyperkahler structure, any divergence-free vector field is a Dirac field. As in this case Ψ is just the volume form in $\mathbf{R}^4$, obviously Dirac fields preserve Ψ.   ⊙

---

[5]And hence also the volume form $\Omega = \Omega^{(1)} \wedge \dots \wedge \Omega^{(n)}$ in $\mathbf{R}^{4n}$.

Chapter 3
Canonical Maps

We have concluded Chap. 2 with the definition of canonical and hyperkahler maps. In this and the next Chapter we will study these two kinds of maps, and their relation to the hyperhamiltonian and related (Dirac) flows. In particular, this Chapter is devoted to canonical transformations, while the next one to hyperkahler transformations.

3.1 Characterization of Canonical Maps

We will now discuss canonical transformations (see Definitions 2.7 and 2.8 of Chap. 2 above); we will denote their set—which quite clearly is actually a group—as $\mathrm{Can}(M)$, or more precisely $\mathrm{Can}(M, g, \mathbf{J})$. We will first discuss the relation between the general case and the case where the structure is in standard form at least at a given point, and then discuss the characterization of canonical maps for structures in standard form. (Later on we will discuss one-parameter groups of canonical transformations, see Sect. 3.2.)

Remark 3.1 As recalled above (see Remark 1.3 in Chap. 1, Sect. 1.4.3), in a previous work [68] we have introduced an operator defined on $(2m \times 2m)$ matrices,

$$\mathcal{P}_{2m}(A) := \frac{1}{2^m \, (m!)} \, \epsilon^{i_1 j_1 \dots i_m j_m} \, A_{i_1 j_1} \dots A_{i_m j_m} \; ; \tag{3.1}$$

this satisfies $[\mathcal{P}_{2m}(A)]^2 = \mathrm{Det}(A)$, i.e. $\mathcal{P}_{2m}(A) = \pm\sqrt{\mathrm{Det}(A)}$. In the case of (4×4) matrices, this is just

$$\mathcal{P}_2(A) = \frac{1}{8} \, \epsilon^{ijkm} \, A_{ij} \, A_{km} \, . \tag{3.2}$$

© Springer International Publishing AG 2017
G. Gaeta and M.A. Rodríguez, *Lectures on Hyperhamiltonian Dynamics and Physical Applications*, Mathematical Physics Studies, DOI 10.1007/978-3-319-54358-1_3

(The operator will also be of use in Sect. 3.5 below, see Remark 3.16.)

Note that in particular for the matrices identifying the standard hypersymplectic structures in $\mathbf{R}^4$ we have

$$\mathcal{P}_2(K_\alpha) = 1, \quad \mathcal{P}_2(\widehat{K}_\alpha) = -1 . \tag{3.3}$$

In this section we will also use a closely related operator, defined on pairs of (4×4) matrices; that is,

$$\mathcal{P}_1(A, B) := \frac{1}{8} \, \epsilon^{ijkm} \, A_{ij} \, B_{km} . \tag{3.4}$$

Note that $\mathcal{P}_1(A, A) = \mathcal{P}_2(A)$. $\odot$

3.1.1 Canonical Maps for Euclidean versus General Manifolds

Let us again fix a reference point $x_0 \in M$ and perform the change of coordinates R which takes the Riemannian metric into standard form at x_0, and subsequently the change of coordinates S which, leaving g in standard form at x_0, takes the complex structures J_α (and hence the symplectic forms ω_α) into standard form at x_0.

Lemma 3.1 *Let $(M, g; \mathbf{J})$ be a hyperkahler manifold; let $R(x)$ be the transformation taking g and J_α into standard form $(g_0 = \delta, \mathbf{J}_0)$ at the point $x \in M$. The group of canonical transformations at the point x is given by*

$$\mathrm{Can}(g, \mathbf{J}) = R^{-1}(x) \, \mathrm{Can}(\delta, \mathbf{J}_0) \, R(x) , \tag{3.5}$$

where $\mathrm{Can}(\delta, \mathbf{J}_0)$ is the group of canonical transformations for g and $\mathbf{J}$ in standard form.

Proof This follows immediately from the tensorial properties of the J_α. In fact, let Y and Y_0 be the matrix representation of J and J_0 respectively; these satisfy $Y = R^{-1}Y_0 R$. Then $A \in \mathrm{Can}(\delta, \mathbf{J}_0)$ means $AY_0 A^{-1} = Y_0$. It follows that for $B = R^{-1}AR$ we have

$$BYB^{-1} = (R^{-1}AR) \, Y \, (R^{-1}A^{-1}R) = R^{-1}A(RYR^{-1})A^{-1}R$$
$$= R^{-1}AY_0 A^{-1}R = R^{-1}Y_0 R = Y .$$

Conversely, if $BYB^{-1} = Y$ this means

$$B \, (R^{-1}Y_0 R) \, B^{-1} = R^{-1}Y_0 R ,$$

and hence

$$(RBR^{-1}) Y_0 (RB^{-1}R^{-1}) = Y_0 .$$

This shows that $(RBR^{-1}) \in \text{Can}(\delta, \mathbf{J}_0)$. We have thus established the isomorphism of $\text{Can}(\delta, \mathbf{J}_0)$ and $\text{Can}(g, \mathbf{J})$ through the map R. △

This Lemma shows that in order to characterize the group of canonical (or strongly canonical) maps for $(g, \mathbf{J})$ we just have to characterize the corresponding group for $(\delta, \mathbf{J}_0)$.

3.1.2 Characterization for Structures in Standard Form

We have then to characterize the group $\text{Can}_0 := \text{Can}(\delta, \mathbf{J}_0)$ of maps which preserve $\iota_{(k)}(\omega \wedge \omega)$ (or $\omega \wedge \omega$ in the case of strongly canonical maps) for standard hyperkahler structures.

The key observation is that[1] for any symplectic form ω we can write the form $\Omega^{(a)}$ built in Sect. 2.7 as

$$\Omega_{(a)} = \pm \iota_a^* [(1/2)(\omega \wedge \omega)] , \tag{3.6}$$

the sign depending on the orientation of $\iota_a^* \omega$.

In local coordinates we have $\omega = K_{ij} dx^i dx^j$ and hence

$$\frac{1}{2} (\omega \wedge \omega) = \frac{1}{2} K_{ij} K_{\ell m} dx^i \wedge dx^j \wedge dx^\ell \wedge dx^m . \tag{3.7}$$

Under a map with Jacobian Λ, K is transformed into $\widetilde{K} = \Lambda^T K \Lambda$; correspondingly, the form $(1/2)(\omega \wedge \omega)$ is rewritten as

$$\frac{1}{2}(\widetilde{\omega} \wedge \widetilde{\omega}) = \frac{1}{2} (\widetilde{K}_{ij} \widetilde{K}_{\ell m}) dx^i \wedge dx^j \wedge dx^\ell \wedge dx^m . \tag{3.8}$$

When we look at the form $\Omega^{(a)}$, only coordinates i, j, ℓ, m in the range $\mathcal{R}_a := [4(a-1)+1, \ldots, 4a]$ should appear; in other words, the operation ι_a^* sets to zero all four-forms $dx^i \wedge dx^j \wedge dx^\ell \wedge dx^m$ except those with exactly (any permutation of) the four suitable coordinates.

Thus we have, with $i, j, \ell, m \in \mathcal{R}_a$,

$$V_{(a)} = \frac{1}{2} (\varepsilon_{ij\ell m} K_{ij} K_{\ell m}) \Omega_{(a)} ; \quad \widetilde{V}_{(a)} = \frac{1}{2} (\varepsilon_{ij\ell m} \widetilde{K}_{ij} \widetilde{K}_{\ell m}) \Omega_{(a)} .$$

The central object is thus the quantity

---

[1] In the same way as, for any symplectic form ω, the volume form Ω on the $4n$-dimensional manifold M can be written as $\Omega = \pm [(1/(2n!))(\omega \wedge \ldots \wedge \omega)]$.

$$\mathcal{P}_2(K) := (1/8) \, (\varepsilon_{ij\ell m} K_{ij} K_{\ell m}) \tag{3.9}$$

(see Sect. 1.4.3, and Remark 3.1); and a map is canonical if and only if

$$\sum_a \iota_a^*[\mathcal{P}_2(\tilde{K})] \equiv \sum_a \iota_a^*[\mathcal{P}_2(\Lambda^T K \Lambda)] = \sum_a \iota_a^*[\mathcal{P}_2(K)] \tag{3.10}$$

for any K corresponding to a symplectic form $\omega \in \mathcal{S}$.

It should be noted that—as easy to check, e.g. by direct computation (see e.g. [68], in particular Appendix A there)—for a generic antisymmetric matrix K it results

$$\mathcal{P}_2(\Lambda^T K \Lambda) = \mathcal{P}_2(K) \, \mathrm{Det}(\Lambda) \, . \tag{3.11}$$

It will again be convenient to write the matrix Λ, as well as the $K = K_\alpha$, in terms of four-dimensional blocks; we will write

$$\Lambda = \begin{pmatrix} A_{11} & \dots & A_{1n} \\ \vdots & \ddots & \vdots \\ A_{n1} & \dots & A_{nn} \end{pmatrix} ; \quad K = \begin{pmatrix} K_{11} & \dots & K_{1n} \\ \vdots & \ddots & \vdots \\ K_{n1} & \dots & K_{nn} \end{pmatrix} , \tag{3.12}$$

with A_{ij} and K_{ij} four-dimensional matrices.

Lemma 3.2 *The group* Can$_0$ *is the group of all the matrices Λ of the form* (3.12) *which satisfy*

$$\frac{1}{n} \sum_{i=1}^n \left[\sum_m \mathcal{P}_2(A_{mi}^T K_{mm}^0 A_{mi}) \right.$$
$$\left. + 2 \sum_{m,\ell} \mathcal{P}_1 \left(A_{mi}^T K_{mm}^0 A_{mi}, A_{\ell i}^T K_{\ell \ell}^0 A_{\ell i} \right) \right] = 1 \, . \tag{3.13}$$

Proof With the notation just introduced, see (3.12), the transformed forms are identified by

$$\tilde{K}_{ij} = A_{\ell i}^T K_{\ell m}^0 A_{mj} \, ;$$

in particular, using $K_{ij}^0 = 0$ for $i \neq j$ (no sum on i),

$$\tilde{K}_{ii} = \sum_m A_{mi}^T K_{mm}^0 A_{mi} \, .$$

Thus, the condition to have a canonical transformation (3.10) reads now

$$\sum_i \mathcal{P}_2 \left[\sum_m A_{mi}^T K_{mm}^0 A_{mi} \right] = \sum_i \mathcal{P}_2 \left[K_{mm}^0 \right] \, ; \tag{3.14}$$

using moreover $P_2[K^0_{mm}] = 1$, this condition is also written as

$$\frac{1}{n} \sum_i P_2 \left[\sum_m A^T_{mi} K^0_{mm} A_{mi} \right] = 1 . \tag{3.15}$$

This can be written in a slightly different form by introducing the notation

$$P_1(A, B) = (1/8) \, \epsilon_{ijkm} \, A_{ij} \, B_{km} ; \tag{3.16}$$

it is easily checked that

$$P_1(B, A) = P_1(A, B) ; \quad P_2(A) = P_1(A, A) . \tag{3.17}$$

Moreover,

$$\begin{aligned} P_2(A + B) &= P_1(A, A) + P_1(A, B) + P_1(B, A) + P_1(B, B) \\ &= P_2(A) + P_2(B) + 2 P_1(A, B) . \end{aligned} \tag{3.18}$$

Using this, condition (3.15) can be rewritten precisely in the form (3.13). This concludes the proof. △

3.2 Canonical Property of the Hyperhamiltonian Flow

We want now to show that the hyperhamiltonian vector fields provide an unfolding of canonical transformations, pretty much in the same way as Hamiltonian vector fields in the case of symplectic structures.

More precisely, we have the following result:

Theorem 3.1 *Let $(M, g; \mathbf{J})$ be a hyperkahler manifold, and let X be any hyperhamiltonian vector field on it w.r.t. the hyperkahler structure $\mathbf{J}$. Then X generates a one-parameter group of canonical transformations in the sense of Definition 2.7 in Chap. 2.*

Remark 3.2 In view of Definition 2.7 of Chap. 2, we have to show that if φ is the flow generated by X, then necessarily

$$\iota^*_a \left[\varphi^*(\omega \wedge \omega) \right] = \iota^*_a(\omega \wedge \omega) \quad \forall \omega \in \mathcal{S}, \ a = 1, ..., n .$$

Our proof will go this way, and be presented after some preliminary results; these will be presented as Lemmas. ⊙

3.2.1 Preliminary Results

We will write $X = X_1 + X_2 + X_3$, with X_α satisfying $X_\alpha \lrcorner \omega_\alpha = d\mathcal{H}^\alpha$ (no sum on α); in coordinates, we have $X_\alpha = f_\alpha^i \partial_i$, with $f_\alpha^i = (M_\alpha)^{ij} \partial_j \mathcal{H}^\alpha$ (again no sum on α).

We will work in local coordinates around the reference point x_0 at which the metric and the hypersymplectic structure are in standard form.[2]

We freely move the indices $\alpha, \beta, \ldots$ (referring to the hyperkahler triple) up and down for typographical convenience.

In order to avoid an exceedingly long proof, we will prove a couple of formulas needed for our argument as separate lemmas.

Lemma 3.3 *For X and ω_α as in Theorem 3.1, it results*

$$\mathcal{L}_X(\omega_\alpha) = \epsilon_{\alpha\beta\gamma} [\partial_i (P_k^\beta (Y_\gamma)_j^k)] dx^i \wedge dx^j . \tag{3.19}$$

Proof We will use the shorthand notations (note $D_\alpha = D_\alpha^T$)

$$P_k^\beta := (\partial H^\beta / \partial x^k) ; \quad D_{ij}^\alpha := \frac{\partial^2 H^\alpha}{\partial x^i \partial x^j} . \tag{3.20}$$

Recall that by definition ω is closed, hence

$$\mathcal{L}_X \omega = d(X \lrcorner \omega) . \tag{3.21}$$

We immediately have, for any $X = f^i \partial_i$, that

$$X \lrcorner \omega_\alpha = \frac{1}{2} \left(K_{ij}^\alpha f^i dx^j - K_{ij}^\alpha f^j dx^i \right) ;$$

rearranging indices, using $K_\alpha^T = -K_\alpha$, and defining

$$\lambda_j^\alpha = f^i (K^\alpha)_{ij} , \tag{3.22}$$

this is also rewritten as

$$X \lrcorner \omega_\alpha = f^i K_{ij}^\alpha dx^j := \lambda_j^\alpha dx^j . \tag{3.23}$$

With this, we get

$$\mathcal{L}_X(\omega_\alpha) = d\lambda_j^\alpha \wedge dx^j = (\partial_i \lambda_j^\alpha) dx^i \wedge dx^j . \tag{3.24}$$

---

[2]Note we are *not* requiring that the hypersymplectic structure (and hence the hyperkahler one as well) are in standard form in a neighborhood of this point; which in general is *not* possible. Our formulas will involve derivatives (hence values at different points) but these will be disposed of later on by using the property of the complex structures of being covariantly constant.

So far we have considered a generic vector field X; in the hyperhamiltonian case, with the notation (3.20), the vector field is given by

$$f^m = M_\beta^{mk} \partial_k H^\beta = P_k^\beta (M_\beta^T)^{km} = - P_k^\beta M_\beta^{km} . \tag{3.25}$$

A simple computation shows that in this case

$$\begin{aligned}
\lambda_j^\alpha &= -P_k^\beta M_\beta^{km} K_{mj}^\alpha = - P_k^\beta (Y_\beta)_m^k (Y_\alpha)_j^m \\
&= - P_k^\beta \left[\epsilon_{\beta\alpha\gamma}(Y_\gamma)_j^k - \delta_{\beta\alpha}\delta_j^k \right] \\
&= P_j^\alpha + \epsilon_{\alpha\beta\gamma} P_k^\beta (Y_\gamma)_j^k .
\end{aligned} \tag{3.26}$$

Differentiating (3.26), and recalling that $\partial_i P_j^\alpha = D_{ij}^\alpha$, we get

$$d\lambda_j^\alpha = D_{ij}^\alpha dx^i + \epsilon_{\alpha\beta\gamma} [\partial_i (P_k^\beta (Y_\gamma)_j^k)] dx^i ; \tag{3.27}$$

now we just recall that $\mathcal{L}_X(\omega_\alpha) = d\lambda_j^\alpha \wedge dx^j$ and observe that $D_{ij}^\alpha dx^i \wedge dx^j = 0$ due to $D_\alpha^T = D_\alpha$. This yields (3.19) and hence concludes the proof. $\triangle$

Remark 3.3 It may be noted that for a generic form $\omega = c_\alpha \omega_\alpha \in S$ (hence with $|c| = \sum c_\alpha^2 = 1$), and setting (here and in the following)

$$\Xi^{ij\ell m} := dx^i \wedge dx^j \wedge dx^\ell \wedge dx^m \tag{3.28}$$

for ease of writing, we obtain easily that

$$\mathcal{L}_X(\omega \wedge \omega) = c_\eta \cdot c_\eta \, \epsilon_{\alpha\beta\gamma} \left[\partial_i \left(P_q^\beta (Y_\gamma)_j^q \right) \right] K_{\ell m}^\alpha \, \Xi^{ij\ell m} . \tag{3.29}$$

If $\partial_i Y^\gamma = 0$ (which is verified in the Euclidean case) we get simply

$$\mathcal{L}_X(\omega_\alpha \wedge \omega_\alpha) = \epsilon_{\alpha\beta\gamma} K_{\ell m}^\alpha D_{iq}^\beta (Y_\gamma)_j^q \, \Xi^{ij\ell m} \tag{3.30}$$

(no sum on α). $\odot$

We can now pass to consider the forms $\Omega^{(a)}$. The key observation here is that $\Omega^{(a)} = dx^{4a-3} \wedge dx^{4a-2} \wedge dx^{4a-1} \wedge dx^{4a}$ (no sum on a), can be written as

$$\Omega^{(a)} = \sigma_a(\omega) \, \iota_a^* [(1/2) (\omega \wedge \omega)] \tag{3.31}$$

for any unimodular ω, where $\sigma_a(\omega) = 1$ for $\iota_a^* \omega \in S$ and $\sigma_a(\omega) = -1$ for $\iota_a^* \omega \in \widehat{S}$. (That is, $\sigma_a(\omega) = \mathcal{P}[\iota_a^*(\omega)]$.)

Any symplectic form in four dimensions is written—at the reference point x_0—as the sum of the standard positively and negatively oriented ones (this just follows from Y_i and $\widehat{Y}_i$ being a basis for the set of all the possible antisymmetric matrices in dimension four); thus we may set in full generality

$$\iota_a^*(\omega) \; = \; c_\alpha \, \omega_\alpha \; + \; \widehat{c}_\alpha \, \widehat{\omega}_\alpha \; . \tag{3.32}$$

(Needless to say, for $\omega \in \mathcal{S}$ only the c_α are nonzero, and conversely for $\widehat{\omega} \in \widehat{\mathcal{S}}$.)

Lemma 3.4 *For X and $\omega \in \Lambda^2(\mathbf{R}^4)$ as in the statement of Theorem 3.1 (thus in particular $\mathrm{d}\omega = 0$),*

$$\begin{aligned}
\mathcal{L}_X(\omega \wedge \omega) = (c_\eta \cdot c_\eta) \, \epsilon_{\alpha\beta\gamma} \left[K^\alpha_{\ell m} \left(D^\beta_{iq}(Y_\gamma)^q{}_j \right. \right. \\
\left. \left. + \, P^\beta_q((A_i)^q{}_m (Y_\gamma)^m{}_j - (Y_\gamma)^q{}_m (A_i)^m{}_j)) \right] \, \Xi^{ij\ell m} \; .
\end{aligned} \tag{3.33}$$

Proof It follows from explicit computations that for such ω,

$$\iota_a^*(\omega \wedge \omega) \; = \; \iota_a^* \left[\sum_\alpha \left[c_\alpha^2 \, (\omega_\alpha \wedge \omega_\alpha) \; + \; \widehat{c}_\alpha^2 \, (\widehat{\omega}_\alpha \wedge \widehat{\omega}_\alpha) \right] \right] , \tag{3.34}$$

with exactly the same c_α and $\widehat{c}_\alpha$ as above; in fact, it is easy to check that $\omega_\alpha \wedge \omega_\beta = 0$ for $\alpha \neq \beta$, and that $\omega_\alpha \wedge \widehat{\omega}_\beta = 0$ for all α and β. We also recall that for symplectic forms (or complex structures) in standard form, all the matrix elements K_{ij} (or $Y^i{}_j$) with i and j not belonging to the same four-dimensional block are zero.

Equation (3.19) yields (no sum on α)

$$\mathcal{L}_X(\omega_\alpha \wedge \omega_\alpha) = \epsilon_{\alpha\beta\gamma} \, K^\alpha_{\ell m} \left(D^\beta_{iq}(Y_\gamma)^q{}_j + P^\beta_q \partial_i (Y_\gamma)^q{}_j \right) \, \Xi^{ij\ell m} \; . \tag{3.35}$$

For a general $\omega = c_\eta \omega_\eta$, this provides

$$\mathcal{L}_X(\omega \wedge \omega) \; = \; (c_\eta \cdot c_\eta) \, \epsilon_{\alpha\beta\gamma} \, K^\alpha_{\ell m} \left(D^\beta_{iq}(Y_\gamma)^q{}_j + P^\beta_q \partial_i (Y_\gamma)^q{}_j \right) \, \Xi^{ij\ell m} \; . \tag{3.36}$$

Here everything can be computed by evaluating matrices *at the single reference point* x_0, except the derivative $\partial_i (Y_\gamma)^q{}_j$. However, this can also be transformed into an algebraic quantity by recalling that, by definition,

$$\nabla J_\gamma \; = \; 0 \; . \tag{3.37}$$

For doing this, we need the connection matrix A; as well known, this is defined in coordinates by

$$(A_i)^j{}_k \; = \; \Gamma^j_{ik} \; , \tag{3.38}$$

with $\Gamma^j_{ik} = \Gamma^j_{ki}$ the Christoffel symbols for the metric g.

Then the condition (3.37) is written in coordinates as

$$\partial_i Y_\gamma \; + \; [A_i, Y_\gamma] \; = \; 0 \; ; \tag{3.39}$$

using (3.39) allows to rewrite (3.36) precisely as (3.33). This concludes the proof of Lemma 3.4.   △

Remark 3.4 Note that in all these formulas, the action of ι_a^* amounts to setting to zero all variables (and its differential) not belonging to the a-th block.   ⊙

3.2.2 Proof of the Main Result

We are now ready to complete our computations, and prove Theorem 3.1; we find it worth discussing first the special (and simpler) case of Euclidean hyperkahler manifolds, which we do in the form of a separate Lemma.

Lemma 3.5 *Theorem 3.1 holds in the Euclidean case.*

Proof In this case (for all γ and all i) we have $\partial_i Y_\gamma = 0$; then (3.36) reduces to

$$\mathcal{L}_X(\omega \wedge \omega) = (c_\eta \cdot c_\eta)\, \epsilon_{\alpha\beta\gamma}\, K^\alpha_{\ell m} D^\beta_{iq}(Y_\gamma)^q{}_j\, dx^i \wedge dx^j \wedge dx^\ell \wedge dx^m . \tag{3.40}$$

It suffices to write down this, or more precisely its pullback under ι_a^*, in explicit terms; for ease of notation we will consider the case $a = 1$, so that only variables $\{x^1, \ldots, x^4\}$ are nonzero (any function should be considered as evaluated with $x^k = 0$ for $k > 4$). We get

$$(1/2)\mathcal{L}_X (\omega \wedge \omega) =$$
$$= \{c_1^2 \left[\left(D_{14}^2 + D_{23}^2 - D_{32}^2 - D_{41}^2\right) + \left(D_{13}^3 - D_{24}^3 - D_{31}^3 + D_{42}^3\right)\right]$$
$$+ c_2^2 \left[\left(D_{12}^1 - D_{21}^1 + D_{34}^1 - D_{43}^1\right) + \left(D_{13}^3 - D_{24}^3 - D_{31}^3 + D_{42}^3\right)\right]$$
$$+ c_3^2 \left[\left(D_{12}^1 - D_{21}^1 + D_{34}^1 - D_{43}^1\right) + \left(D_{14}^2 + D_{23}^2 - D_{32}^2 - D_{41}^2\right)\right]\} \times$$
$$\times dx^1 \wedge dx^2 \wedge dx^3 \wedge dx^4 .$$

Recalling that $D_{ij}^\alpha = D_{ji}^\alpha$ we conclude that each of the coefficients of the c_α^2 vanish separately, hence $\mathcal{L}_X(\omega \wedge \omega) = 0$ as stated.   △

We will now come to consider the general case[3] and complete our proof.

Proof of Theorem 3.1 The variation of $\iota_a^*(\omega \wedge \omega)$ under X is given by $\iota_a^*[\mathcal{L}_X(\omega \wedge \omega)]$. To evaluate this we make use of (3.33) and of Remark 3.4. Moreover, for ease of notation, we will focus on $a = 1$; that is, we should compute (3.33) with all i, k, ℓ, m indices restricted to the range $1, \ldots, 4$.

We note that according to (3.33), $\mathcal{L}_X(\omega \wedge \omega)$, and therefore $\iota_a^*[\mathcal{L}_X(\omega \wedge \omega)]$ as well, is the sum of two terms; these correspond respectively to $K^\alpha_{\ell m} D^\beta_{iq}(Y_\gamma)^q{}_j$ and to

---

[3]Or at least as general as allowed by Remark 2.12 in Chap. 2; see the next Sect. 3.2.3 for a discussion.

$K^\alpha_{\ell m} P^\beta_q [A_i, Y_\gamma]^q_j$. The first term is exactly the one which was already evaluated in the case $\partial_i Y_\gamma = 0$; it vanishes as stated by Lemma 3.5 (and shown in its proof).

Therefore, defining

$$\Theta^{(a)}_{\alpha\beta\gamma} = \iota^*_a \left[K^\alpha_{\ell m} P^\beta_q \left((A_i)^q_{\ p} (Y_\gamma)^p_{\ j} - (Y_\gamma)^q_{\ p} (A_i)^p_{\ j} \right) \Xi^{ij\ell m} \right] ,$$

we only have to show that

$$\frac{c_\alpha}{2} \epsilon_{\alpha\beta\gamma} \Theta^{(a)}_{\alpha\beta\gamma} = 0 . \qquad (3.41)$$

Note that here the coefficients c_α (satisfying $|c|^2 = 1$) and the vectors P^β are completely arbitrary; thus (3.41) should hold for any choice of these. In other words we should have

$$\Theta^{(a)}_{\alpha\beta\gamma} = 0 \qquad (3.42)$$

for all choices of the indices α, β, γ, provided these are all different, $\alpha \neq \beta \neq \gamma \neq \alpha$.

The expression for Θ only involves quantities computed at the reference point x_0, and we can hence make use of the explicit expressions for the standard form of the Y_α and the K_α.

Using these, choosing $a = 1$ (and omitting the index a) for ease of notation, and the case of positive orientation in the first block, we get e.g.

$$\begin{aligned}
\Theta_{123} = -2 \{&[(\Gamma^1_{14} + \Gamma^1_{23} - \Gamma^1_{32} - \Gamma^1_{41}) - (\Gamma^3_{12} - \Gamma^3_{21} + \Gamma^3_{34} - \Gamma^3_{43})] \, P^2_1 \\
&+ [(\Gamma^2_{14} + \Gamma^2_{23} - \Gamma^2_{32} - \Gamma^2_{41}) + (\Gamma^4_{12} - \Gamma^4_{21} + \Gamma^4_{34} - \Gamma^4_{43})] \, P^2_2 \\
&+ [(\Gamma^1_{12} - \Gamma^1_{21} + \Gamma^1_{34} - \Gamma^1_{43}) + (\Gamma^3_{14} + \Gamma^3_{23} - \Gamma^3_{32} - \Gamma^3_{41})] \, P^2_3 \\
&+ [(\Gamma^2_{21} - \Gamma^2_{12} + \Gamma^2_{43} - \Gamma^2_{34}) + (\Gamma^4_{14} + \Gamma^4_{23} - \Gamma^4_{32} - \Gamma^4_{41})] \, P^2_4 \} \times \\
&\times \mathrm{d}x^1 \wedge \mathrm{d}x^2 \wedge \mathrm{d}x^3 \wedge \mathrm{d}x^4 ;
\end{aligned}$$

recalling the property $\Gamma^i_{jk} = \Gamma^i_{kj}$, valid for any torsion-free Riemannian metric g, it is immediately seen that Θ_{123} vanishes. The same holds for all forms $\Theta_{\alpha\beta\gamma}$ with $\alpha \neq \beta \neq \gamma \neq \alpha$ (explicit formulas are omitted for the sake of brevity).

This concludes the proof for positive orientation. As usual, computations are the same up to certain signs for negative orientation as well, and lead to the same result.

We have thus shown that a hyperhamiltonian flow preserves $\iota^*_a(\omega \wedge \omega)$, and hence the forms $\Omega^{(a)}$. Theorem 3.1 follows immediately, see Remark 3.2.   △

3.2.3 Canonical Maps and Holonomy

Thanks to the use of Eq. (3.39), our discussion was based at a single reference point $x_0 \in M$ and hence could use the standard form of the hyperkahler structure; and this also for non-Euclidean metrics. It should however be recalled, see Remark 2.12 in

Chap. 2, that if the holonomy group for (the Levi-Civita connection ∇ associated to) the given metric is non-trivial, or at least is not contained in $\mathrm{Sp}(1) \times \ldots \times \mathrm{Sp}(1) \subset \mathrm{Sp}(n)$, the very splitting of $T_{x_0} M \simeq \mathbf{R}^{4n}$ as

$$T_{x_0} M = \mathbf{R}^4 \oplus \ldots \oplus \mathbf{R}^4$$

is ill-defined.

In other words, in this case it is true that the we have canonical maps in the sense of Definition 2.7 (Chap. 2) for the given splitting of $T_{x_0} M$, but this splitting is not invariant under the holonomy group (that is, under the Levi-Civita connection.

On the other hand, the action of the holonomy group on this splitting will in general produce a different splitting; but our discussion can be applied to this one too.

We believe it is useful discussing this matter in coordinates; let us consider the case $n = 2$ to fix ideas (the general n case is analogous), and use a four-dimensional block notation.

Thus the complex structures will be defined (in the splitting coordinates) by block matrices

$$Y_\alpha = \begin{pmatrix} Y_\alpha^{(1)} & 0 \\ 0 & Y_\alpha^{(2)} \end{pmatrix}$$

and the holonomy group action will be through matrices

$$G = \begin{pmatrix} A & B \\ C & D \end{pmatrix},$$

where each of the $A, \ldots, D$ entries is actually an $\mathrm{SU}(2)$ matrix. A vector $\mathbf{v} \in T_{x_0} M$ will be written as

$$\mathbf{v} = \begin{pmatrix} \xi^{(1)} \\ \xi^{(2)} \end{pmatrix}, \quad \xi^{(i)} \in \mathbf{R}^4.$$

It is clear that vectors in $\mathbf{R}_{(1)}^4$, i.e. with $\xi^{(2)} = 0$, will be mapped into

$$\widetilde{\mathbf{v}} = \begin{pmatrix} \widetilde{\xi}^{(1)} \\ \widetilde{\xi}^{(2)} \end{pmatrix} = \begin{pmatrix} A \, \xi^{(1)} \\ C \, \xi^{(1)} \end{pmatrix},$$

thus characterized by

$$A^{-1} \widetilde{\xi}^{(1)} = C^{-1} \widetilde{\xi}^{(2)} ;$$

and correspondingly vectors in $\mathbf{R}_{(2)}^4$, i.e. with $\xi^{(1)} = 0$ will be mapped into

$$\widetilde{\mathbf{v}} = \begin{pmatrix} \widetilde{\xi}^{(1)} \\ \widetilde{\xi}^{(2)} \end{pmatrix} = \begin{pmatrix} B \, \xi^{(2)} \\ D \, \xi^{(2)} \end{pmatrix},$$

thus characterized by

$$B^{-1}\widetilde{\xi}^{(1)} \;=\; D^{-1}\widetilde{\xi}^{(2)}\,.$$

That is, albeit there is no way to define a splitting which holds for *all* elements G of the holonomy group $\mathrm{Sp}(2)$, for *each* element G of the holonomy group a splitting is easily determined—actually, just by acting with G^{-1}, i.e. going back to the original splitting.

In other words, if the element g of the holonomy group maps $\mathbf{J}_0$ into $\widetilde{\mathbf{J}} = g\mathbf{J}_0 g^{-1}$, and χ is canonical for $\mathbf{J}_0$, then $\widetilde{\chi} = g\chi g^{-1}$ will be canonical for $\widetilde{\mathbf{J}}$. This is just a different interpretation of the discussion provided in Sect. 3.1.

3.2.4   Canonical Property of Dirac Vector Fields

We have shown that the hyperhamiltonian flow is canonical for the underlying hyperkahler structure; it turns out it is also canonical for any dual one.

Indeed, the structures ω_α and $\widehat{\omega}_\alpha$ define the same invariant subspaces in $T_x M$ and define on these volume forms which only differ by a sign, $\Omega^{(a)} = -\widehat{\Omega}^{(a)}$; it is thus a triviality that preservation of the forms $\Omega^{(a)}$ for the defining structure entails preservation of the $\Omega^{(a)}$ for any dual one. In other words, $\mathcal{L}_X(\omega \wedge \omega) = 0$ implies $\mathcal{L}_X(\widehat{\omega} \wedge \widehat{\omega}) = 0$; this follows at once from $(\widehat{\omega} \wedge \widehat{\omega}) = -(\omega \wedge \omega)$.

Reversing the roles of the vector field and of the symplectic structures, this shows that any Dirac vector field is canonical, i.e. the generator of a one-parameter group of canonical maps.

We state this formally as:

Theorem 3.2 *Let $(M, g; \mathbf{J})$ be a hyperkahler manifold, $\widehat{\mathbf{J}}$ any hyperkahler structure dual to $\mathbf{J}$, and let X be any vector field on it which is hyperhamiltonian w.r.t. the hyperkahler structure $\widehat{\mathbf{J}}$ (i.e. X is a Dirac vector field for $\mathbf{J}$). Then X generates a one-parameter group of canonical transformations in the sense of Definition 2.7 in Chap. 2.*

Remark 3.5 It may be worth recalling that dual hyperkahler structures in $\mathbf{R}^{4n}$ have hyperkahler forms which are not related by any simple operation such as a global sign switch (see Sect. 2.6), except for $n = 1$. Thus (for $n > 1$) a vector field which generates a group of hyperkahler maps for a given hyperkahler structure will in general *not* generate hyperkahler maps for the dual structures. This is quite different from the behavior just described in terms of canonical transformations. The reason can be expressed in quite simple terms: in the case of canonical transformations we are only interested in the restriction of the hyperkahler form $\Psi = (1/2)\omega_\alpha \wedge \omega_\alpha$ to the irreducible hyperkahler components of TM, i.e. disregard all terms in $\Lambda^2(M)$ which "bridge" different blocks.   ⊙

3.2.5 Canonical Maps in Euclidean $\mathbf{R}^4$

Let us now consider in some detail $(\mathbf{R}^4, \delta)$ with standard hyperkahler structure (with positive orientation) from the point of view of canonical maps. The volume form is now just

$$\Omega = dx^1 \wedge dx^2 \wedge dx^3 \wedge dx^4 .$$

We consider an arbitrary $\omega \in \mathcal{S}$, i.e. $\omega = c_\alpha \omega_\alpha$ with $|c|^2 = c_1^2 + c_2^2 + c_3^2 = 1$; for this we have $(1/2)(\omega \wedge \omega) = \Omega$. For a vector field $X = f^i \partial_i$ it follows from standard computations (using also $|c|^2 = 1$) that

$$(X \lrcorner \omega) \wedge \omega = f^1 dx^2 \wedge dx^3 \wedge dx^4 - f^2 dx^1 \wedge dx^3 \wedge dx^4$$
$$+ f^3 dx^1 \wedge dx^2 \wedge dx^4 - f^4 dx^1 \wedge dx^2 \wedge dx^3 . \quad (3.43)$$

Specifying now that X is the hyperhamiltonian vector field corresponding to Hamiltonians $\{\mathcal{H}_1, \mathcal{H}_2, \mathcal{H}_3\}$, we get

$$(X \lrcorner \omega) \wedge \omega = (\partial_2 \mathcal{H}_1 + \partial_4 \mathcal{H}_2 + \partial_3 \mathcal{H}_3) \, dx^2 \wedge dx^3 \wedge dx^4$$
$$+ (\partial_1 \mathcal{H}_1 - \partial_3 \mathcal{H}_2 + \partial_4 \mathcal{H}_3) \, dx^1 \wedge dx^3 \wedge dx^4$$
$$+ (\partial_4 \mathcal{H}_1 - \partial_2 \mathcal{H}_2 - \partial_1 \mathcal{H}_3) \, dx^1 \wedge dx^2 \wedge dx^4$$
$$+ (\partial_3 \mathcal{H}_1 + \partial_1 \mathcal{H}_2 - \partial_2 \mathcal{H}_3) \, dx^1 \wedge dx^2 \wedge dx^3 . \quad (3.44)$$

It follows from this that

$$\mathcal{L}_X(\omega \wedge \omega) = d[(X \lrcorner \omega) \wedge \omega] = 0 . \quad (3.45)$$

In other words, we have shown by explicit computation that

$$\Omega = (1/2) \, (\omega \wedge \omega)$$

is preserved under *any* hyperhamiltonian flow. (Actually our computation showed this only for positively oriented hypersymplectic structures; the computation goes the same way for negatively oriented ones.)

Let us go back to considering $\mathcal{L}_X(\omega)$; using the explicit expression for the hyperhamiltonian vector field, it turns out by a direct computation that this can be written as

$$\mathcal{L}_X(\omega) = \frac{1}{2} \, (p_\alpha \, \omega_\alpha + q_\alpha \, \widehat{\omega}_\alpha) \quad (3.46)$$

with coefficients p_α, q_α given by (here $\triangle$ is the Laplacian)

$$p_1 = c_2 \Delta \mathcal{H}_3 - c_3 \Delta \mathcal{H}_2 \,, \quad p_2 = c_3 \Delta \mathcal{H}_1 - c_1 \Delta \mathcal{H}_3 \,, \quad p_3 = c_1 \Delta \mathcal{H}_2 - c_2 \Delta \mathcal{H}_1 \,;$$

$$\begin{aligned}
q_1 &= c_1 \left[\left(\partial_1^2 \mathcal{H}_2 - \partial_2^2 \mathcal{H}_2 + \partial_3^2 \mathcal{H}_2 - \partial_4^2 \mathcal{H}_2 \right) - 2 \left(\partial_1 \partial_2 \mathcal{H}_3 + \partial_3 \partial_4 \mathcal{H}_3 \right) \right] \\
&\quad - c_2 \left[\left(\partial_1^2 \mathcal{H}_1 - \partial_2^2 \mathcal{H}_1 + \partial_3^2 \mathcal{H}_1 - \partial_4^2 \mathcal{H}_1 \right) - 2 \left(\partial_1 \partial_4 \mathcal{H}_3 - \partial_2 \partial_3 \mathcal{H}_3 \right) \right] \\
&\quad + 2 c_3 \left[\left(\partial_1 \partial_2 \mathcal{H}_1 + \partial_3 \partial_4 \mathcal{H}_1 \right) + \left(\partial_1 \partial_4 \mathcal{H}_2 - \partial_2 \partial_3 \mathcal{H}_2 \right) \right] \,,
\end{aligned}$$

$$\begin{aligned}
q_2 &= c_1 \left[\left(\partial_1^2 \mathcal{H}_3 - \partial_2^2 \mathcal{H}_3 - \partial_3^2 \mathcal{H}_3 + \partial_4^2 \mathcal{H}_3 \right) + 2 \left(\partial_1 \partial_2 \mathcal{H}_2 - \partial_3 \partial_4 \mathcal{H}_2 \right) \right] \\
&\quad - c_3 \left[\left(\partial_1^2 \mathcal{H}_1 - \partial_2^2 \mathcal{H}_1 - \partial_3^2 \mathcal{H}_1 + \partial_4^2 \mathcal{H}_1 \right) - 2 \left(\partial_1 \partial_3 \mathcal{H}_2 + \partial_2 \partial_4 \mathcal{H}_2 \right) \right] \\
&\quad - 2 c_2 \left[\left(\partial_1 \partial_2 \mathcal{H}_1 - \partial_3 \partial_4 \mathcal{H}_1 \right) + \left(\partial_1 \partial_3 \mathcal{H}_3 + \partial_2 \partial_4 \mathcal{H}_3 \right) \right] \,,
\end{aligned}$$

$$\begin{aligned}
q_3 &= c_2 \left[\left(-\partial_1^2 \mathcal{H}_3 - \partial_2^2 \mathcal{H}_3 + \partial_3^2 \mathcal{H}_3 + \partial_4^2 \mathcal{H}_3 \right) + 2 \left(\partial_1 \partial_4 \mathcal{H}_1 + \partial_2 \partial_3 \mathcal{H}_1 \right) \right] \\
&\quad + c_3 \left[\left(\partial_1^2 \mathcal{H}_2 + \partial_2^2 \mathcal{H}_2 - \partial_3^2 \mathcal{H}_2 - \partial_4^2 \mathcal{H}_2 \right) + 2 \left(\partial_1 \partial_3 \mathcal{H}_1 - \partial_2 \partial_4 \mathcal{H}_1 \right) \right] \\
&\quad - 2 c_1 \left[\left(\partial_1 \partial_4 \mathcal{H}_2 + \partial_2 \partial_3 \mathcal{H}_2 \right) - \left(\partial_1 \partial_3 \mathcal{H}_3 - \partial_2 \partial_4 \mathcal{H}_3 \right) \right] \,.
\end{aligned}$$

The essential point here is that—as these explicit formulas show—the Lie derivative $\mathcal{L}_X(\omega)$ of a symplectic form $\omega \in \mathcal{S} \subset \mathcal{Q}$ has components along $\hat{\mathcal{Q}}$, i.e. the negatively-oriented forms.

This shows that in general the hyperhamiltonian flow, even in this simple case, is canonical but *not* hyperkahler; see also Remark 2.11 in Chap. 2.

An exception is provided by the choice $\mathcal{H}_1 = \mathcal{H}_2 = \mathcal{H}_3 = (1/2)(x_1^2 + x_2^2 + x_3^2 + x_4^2)$, corresponding to the "quaternionic oscillator" (which is an integrable case [59, 61]) to be discussed in Chap. 5. With this, we get

$$p_1 = 4(c_2 - c_3), \quad p_2 = 4(c_3 - c_1), \quad p_3 = 4(c_1 - c_2); \quad q_1 = q_2 = q_3 = 0 \,.$$

It is also worth mentioning what happens when only one of the Hamiltonians, say $\mathcal{H}_1$, is nonzero; this corresponds to a standard Hamiltonian flow. Setting $\mathcal{H}_1 = H$, $\mathcal{H}_2 = \mathcal{H}_3 = 0$ in the general formulas above, we get

$$\begin{aligned}
p_1 &= 0 \,, \quad p_2 = c_3 \Delta H \,, \quad p_3 = -c_2 \Delta H \,; \\
q_1 &= -c_2 \left(\partial_1^2 H - \partial_2^2 H + \partial_3^2 H - \partial_4^2 H \right) + 2 c_3 \left(\partial_1 \partial_2 H + \partial_3 \partial_4 H \right) \,, \\
q_2 &= -c_3 \left(\partial_1^2 H - \partial_2^2 H - \partial_3^2 H + \partial_4^2 H \right) - 2 c_2 \left(\partial_1 \partial_2 H \partial_3 \partial_4 H \right) \,, \\
q_3 &= 2 c_2 \left(\partial_1 \partial_4 H + \partial_2 \partial_3 H \right) + 2 c_3 \left(\partial_1 \partial_3 H - \partial_2 \partial_4 H \right) \,.
\end{aligned}$$

This shows that—as already remarked above—even a simple Hamiltonian flow is in general hyperkahler but *not* strongly hyperkahler.[4]

---

[4]Note in [68] there is an imprecise statement in this respect.

3.3 The Inverse Problem: General One-Parameter Groups of Canonical Transformations in $\mathbf{R}^{4n}$

We have shown (see Sect. 3.2) that any hyperhamiltonian flow provides a one-parameter group of canonical transformations for the underlying hyperkahler structure. We have also shown that this property is actually shared by any Dirac vector field (for the same hyperkahler structure).

Here we want to discuss what is the most general vector field which generates a one-parameter group of canonical transformations for a given hyperkahler structure. It will turn out there is nothing new, i.e. these are the Dirac vector fields for the given structure. Note this generalizes a well known result in Hamiltonian dynamics [10, 30, 82, 109, 111].

Note that in this Section we only deal with $\mathbf{R}^{4n}$ Euclidean spaces; the same kind of considerations presented in Sects. 2.7.3 and 3.1 would apply when trying to extend these results to general manifolds, in particular to manifolds with nontrivial holonomy group.

As usual, the case where $(M, g) = (\mathbf{R}^4, \delta)$ is special in many ways, and accordingly we will discuss it separately before going over to the general one.

3.3.1 The Case $n = 1$

As mentioned above, the case $n = 1$, i.e. $M = \mathbf{R}^4$ is special in several ways; it is of course also quite simpler than the general one, and we will start by analyzing it. It turns out that—albeit the results obtained in this section are specific to the case $\mathbf{R}^4$, and do not hold for $\mathbf{R}^{4n}$ for $n > 1$—the essential features of our general problem are already present in this case.

We will use coordinates $x^1, \ldots, x^4$. First of all, we recall that the hyperkahler structure can always be reduced to one of the standard ones, either (1.23) or (1.25) depending on its orientation, see Sect. 1.4.3; correspondingly the symplectic structures will be given by either (1.24) or (1.26).

The condition for a vector field

$$Z = f^i(x)\, \partial_i \qquad (3.47)$$

to be canonical is just that it preserves the volume form $\Omega = \omega \wedge \omega$ (with ω any form in the symplectic Kahler sphere). We thus just have to study divergence-free vector fields in $\mathbf{R}^4$.

Lemma 3.6 *Any divergence-free vector field in $\mathbf{R}^4$ can be written as a Dirac vector field, i.e. as the sum of vector fields which are hyperhamiltonian w.r.t. the pair of dual standard hyperkahler structures.*

Proof Denoting by Ω the volume form on $\mathbf{R}^4$, i.e. $\Omega = \mathrm{d}x^1 \wedge \mathrm{d}x^2 \wedge \mathrm{d}x^3 \wedge \mathrm{d}x^4$, we associate to Z the three-form

$$\Phi = Z \lrcorner \Omega \in \Lambda^3(M) ; \qquad (3.48)$$

this is given in coordinates by $\Phi = (1/6)\epsilon_{ijkm} f^i \mathrm{d}x^j \wedge \mathrm{d}x^k \wedge \mathrm{d}x^m$,

$$\begin{aligned}
\Phi = {} & f^1 \mathrm{d}x^2 \wedge \mathrm{d}x^3 \wedge \mathrm{d}x^4 - f^2 \mathrm{d}x^1 \wedge \mathrm{d}x^3 \wedge \mathrm{d}x^4 \\
& + f^3 \mathrm{d}x^1 \wedge \mathrm{d}x^2 \wedge \mathrm{d}x^4 - f^4 \mathrm{d}x^1 \wedge \mathrm{d}x^2 \wedge \mathrm{d}x^3 .
\end{aligned} \qquad (3.49)$$

Thus $\mathrm{div}(f) \cdot \Omega = \mathrm{d}\Phi$, and the condition $\mathrm{div}(f) = 0$ is equivalent to $\mathrm{d}\Phi = 0$. As $H^3(\mathbf{R}^4) = 0$ (and actually for any four-dimensional manifold M with $H^3(M) = 0$), this also implies that $\Phi = \mathrm{d}\Psi$ for some $\Psi \in \Lambda^2(M)$. (Note that this Ψ is not uniquely defined; we will just pick up one specific Ψ, see also Remark 3.8 below).

We can always write such a Ψ in coordinates as

$$\Psi = \frac{1}{2} P_{ij}(x) \, \mathrm{d}x^i \wedge \mathrm{d}x^j \qquad (3.50)$$

for some skew-symmetric matrix function P (identified by six scalar functions P_{ij}, say with $i < j$).

Writing down explicitly $\mathrm{d}\Psi$ in this notation, and comparing with (3.49), we immediately get the expression of f^i in terms of the partial derivatives of the P_{jk}; these are compactly written as

$$f^h = \epsilon^{hk\ell m} \partial_k P_{\ell m} . \qquad (3.51)$$

On the other hand, the most general expression of a vector field $Z = f^i \partial_i = X + \widehat{X}$ as given in the statement is obtained for

$$f^i = \sum_{a=1}^{3} \left[M_a^{ij} (\partial_j \mathcal{H}^a) + \widehat{M}_a^{ij} (\partial_j \widehat{\mathcal{H}}^a) \right] . \qquad (3.52)$$

By looking at the explicit expression for $M_a = Y_a g^{-1}$ and $\widehat{M}_a = \widehat{Y}_a g^{-1}$ (where now the metric is just $g = \delta$) we obtain an explicit expression (not reported here) for the f^i. Comparing this and (3.51) we see that we should have

$$\begin{aligned}
P_{12} = \mathcal{H}^1 + \widehat{\mathcal{H}}^3 , \quad P_{13} = \mathcal{H}^3 - \widehat{\mathcal{H}}^1 , \quad P_{14} = \mathcal{H}^2 + \widehat{\mathcal{H}}^2 , \\
P_{23} = \mathcal{H}^2 - \widehat{\mathcal{H}}^2 , \quad P_{24} = -\mathcal{H}^3 - \widehat{\mathcal{H}}^1 , \quad P_{34} = \mathcal{H}^1 - \widehat{\mathcal{H}}^3 ;
\end{aligned} \qquad (3.53)$$

conversely, we have

$$\mathcal{H}^1 = \frac{P_{12} + P_{34}}{2}, \quad \mathcal{H}^2 = \frac{P_{14} + P_{23}}{2}, \quad \mathcal{H}^3 = \frac{P_{13} - P_{24}}{2};$$

$$\widehat{\mathcal{H}}^1 = -\frac{P_{13} + P_{24}}{2}, \quad \widehat{\mathcal{H}}^2 = \frac{P_{14} - P_{23}}{2}, \quad \widehat{\mathcal{H}}^3 = \frac{P_{12} - P_{34}}{2}. \quad (3.54)$$

This shows that the vector field $Z = f^i \partial_i$ with $\partial_i f^i = 0$ is in fact always expressed as the sum of vector fields X and $\widehat{X}$ which are hyperhamiltonian w.r.t. the standard positively and negatively oriented standard hyperkahler structure in $\mathbf{R}^4$, and thus completes our proof.[5] △

Corollary 3.1 *Any smooth one-parameter family of canonical transformations for a hyperkahler structure in Euclidean $\mathbf{R}^4$ can be written as the flow of a Dirac vector field.*

Remark 3.6 The result of Lemma 3.6 can also be stated in a different way: we have shown that (in Euclidean $\mathbf{R}^4$) if a vector field has zero divergence, then it is necessarily a Dirac vector field. The result is not surprising. In fact, the matrices Y_a and $\widehat{Y}_a$ are a basis for the set of antisymmetric matrices in $\mathbf{R}^4$. ⊙

Remark 3.7 It is well known that, thanks to the classical Hodge theorem, any vector field $X = f^i \partial_i$ in $\mathbf{R}^n$ can be decomposed into a gradient one and one with zero divergence; that is, one can always write

$$X = X_1 + X_2; \quad X_1 = f_1^i \partial_i, \quad X_2 = f_2^i \partial_i,$$

where $f_1^i = g^{ij} \partial_j U$ for some potential U, and $\mathrm{div}(X_2) = \partial_i f_2^i = 0$.

In dimension two, any divergence-free vector field can be written as a Hamiltonian one, and thus Hodge theorem also implies that in dimension two any vector field can be written as the sum of a gradient and a Hamiltonian (w.r.t. the standard symplectic structure) vector fields.

Our Lemma 3.6 above shows that a similar result also holds in dimension four: that is, any vector field can be written as the sum of a gradient and a Dirac (w.r.t. the pair of dual standard hypersymplectic structures) vector fields.

As mentioned in Remark 3.6 above, this just amounts to the fact that the $(3+3)$ matrices Y_a and $\widehat{Y}_a$ are a basis for the (six dimensional) space of antisymmetric matrices in $\mathbf{R}^4$.

A more refined way of stating this fact is as follows: in $\mathbf{R}^4$, the tail end of the De Rham complex can be written as

$$
\begin{array}{c}
\Lambda_+^2 \\
\oplus \quad \searrow \\
\quad \quad \nearrow \quad \Lambda^3 \to \Lambda^4 \\
\Lambda_-^2
\end{array}
$$

---

[5]Note that in order to explicitly determine the $\mathcal{H}^a$, $\widehat{\mathcal{H}}^a$ (or equivalently the P_{ij}) for a given vector field $Z = f^i \partial_i$ we should integrate the Eq. (3.51), or equivalently (3.52); this can be a nontrivial problem.

where $\Lambda^2_\pm$ are the bundles of (anti)-self-dual two forms; this sequence is globally exact on $\mathbf{R}^4$. We thank an anonymous referee to some previous work of ours for pointing out this fact.   ⊙

Remark 3.8 As mentioned above, the two-form Ψ considered in the Proof of Lemma 3.6 and satisfying $\Phi = \mathrm{d}\Psi$ is not uniquely defined; two such forms $\Psi^{(1)}$, $\Psi^{(2)}$ satisfy $\mathrm{d}(\Psi^{(1)} - \Psi^{(2)}) = 0$. Correspondingly, to these correspond matrices $P^{(1)}$, $P^{(2)}$ such that $Q := P^{(1)} - P^{(2)}$ satisfies $\epsilon^{ijkl}(\partial_j Q_{kl}) = 0$.

On the other hand, the $\mathcal{H}^a$ which yield a given HH vector field, i.e. given f^i functions, are also not uniquely defined; this is apparent from (3.52).

If we define the matrix

$$W^{ij} = \sum_{a=1}^{3} \left(M_a^{ij} \mathcal{H}^a + \hat{M}_a^{ij} \widehat{\mathcal{H}}^a \right) , \tag{3.55}$$

we can write

$$f^i = \epsilon^{ijkl} \partial_j W_{kl} . \tag{3.56}$$

Comparing this and (3.51) we see that we have

$$\epsilon^{ijkm} \partial_j (W_{km} - P_{km}) = 0 . \tag{3.57}$$

A particular solution of this equation is the trivial one, $W_{km} = P_{km}$, considered in the Proof above. This yields (3.53) and (3.54). Any other solution reflects the non-uniqueness of P_{ij} and W_{ij}.   ⊙

3.3.2 The General Case

We will now consider $M = \mathbf{R}^{4n}$ with the Euclidean metric; we work in coordinates $x^1, \ldots, x^{4n}$ in $M = \mathbf{R}^{4n}$. As already mentioned, we can always reduce to standard HK structures. In any case, we have a splitting of TM into four-dimensional blocks,

$$T_x M = \mathbf{R}^4_{(1)} \oplus \ldots \oplus \mathbf{R}^4_{(n)} ,$$

and the hyperkahler structure is written as the direct sum of hyperkahler structures (of either orientation) in each block. We recall that ι_k denotes the embedding of $\mathbf{R}^4_{(k)}$ into $\mathbf{R}^{4n}$, and $\omega_\alpha^{(k)} := \iota_k^*(\omega_\alpha)$.

In the following we will just consider the case where the orientation is always positive, for ease of discussion; the argument in the general case goes exactly the same way, *mutatis mutandis*.

We start by introducing a convenient notation. We rewrite the coordinates as $\xi^i_{(k)}$, where $k = 1, \ldots, n$ is the block index, and $i = 1, \ldots, 4$. We choose

$$\xi^i_{(k)} = x^{4(k-1)+i} ,\tag{3.58}$$

so that the coordinates are now

$$(\xi^1_{(1)}, \ldots, \xi^4_{(1)}; \xi^1_{(2)}, \ldots, \xi^4_{(n)}) .\tag{3.59}$$

The symplectic structures are thus written as[6]

$$\omega_1 = \sum_{k=1}^n (d\xi^1_{(k)} \wedge d\xi^2_{(k)} + d\xi^3_{(k)} \wedge d\xi^4_{(k)}) = \sum_{k=1}^n \omega_1^{(k)} ,$$

$$\omega_2 = \sum_{k=1}^n (d\xi^1_{(k)} \wedge d\xi^4_{(k)} + d\xi^2_{(k)} \wedge d\xi^3_{(k)}) = \sum_{k=1}^n \omega_2^{(k)} ,$$

$$\omega_3 = \sum_{k=1}^n (d\xi^1_{(k)} \wedge d\xi^3_{(k)} + d\xi^4_{(k)} \wedge d\xi^2_{(k)}) = \sum_{k=1}^n \omega_3^{(k)} .$$

Now we would like to characterize vector fields which are canonical (see Sect. 2.7), i.e. such that

$$\iota^*_k[\varphi^*(\omega \wedge \omega)] = \iota^*_k(\omega \wedge \omega)\tag{3.60}$$

for all $\omega \in S$ and all k, with φ the flow. These can also be equivalently characterized as the vector fields Z such that, for all ω and all k,

$$\iota^*_k[L_Z(\omega \wedge \omega)] = 0 ,\tag{3.61}$$

with L the Lie derivative.

Theorem 3.3 *The vector field Z is the generator of a one-parameter family of canonical transformations for a hyperkahler structure in the Euclidean $\mathbf{R}^{4n}$ space, i.e. satisfies (3.61) for the associated hypersymplectic structure, if and only if it is a Dirac vector field for the underlying hyperkahler structure.*

Proof We will be writing a generic vector field as $Z = f^i(x)\partial_i$. Moreover, writing as above $\omega = \sum_{k=1}^n \omega^{(k)} = \sum_{k=1}^n \iota^*_k(\omega)$, the form $\omega \wedge \omega$ is written as

$$2\eta := \omega \wedge \omega = \sum_{k,\ell} \omega^{(k)} \wedge \omega^{(\ell)} .\tag{3.62}$$

Note that ι^*_κ acts by selecting *only* the components with block index κ, i.e. (with no sum on κ)

$$\iota^*_\kappa(\omega^{(\ell)} \wedge \omega^{(m)}) = \delta_{\ell\kappa} \delta_{m\kappa} \omega^{(\ell)} \wedge \omega^{(m)} .\tag{3.63}$$

---

[6]If both orientations are present, we would rearrange coordinates so that the first k_1 blocks are positive, and the sum would be from 1 to k_1, with then another sum from $k_1 + 1$ to n with the expressions corresponding to the $\widehat{\omega}_\alpha$.

We also write

$$\omega \; = \; \frac{1}{2} \, A_{ij} \mathrm{d}x^i \wedge \mathrm{d}x^j \; ; \tag{3.64}$$

the matrix A will be written as

$$A \; = \; A^{(1)} \oplus \ldots \oplus A^{(n)} \, , \tag{3.65}$$

where $A^{(\kappa)}$ is the usual four-dimensional block, which we also denote as K. This also means that (no sum on a) $\omega^{(a)} = (1/2) K_{ij} \mathrm{d}\xi^i_{(a)} \wedge \mathrm{d}\xi^j_{(a)}$, and therefore

$$\omega \; = \; \sum_{a=1}^{n} \frac{1}{2} \, K_{ij} \, \mathrm{d}\xi^i_{(a)} \wedge \mathrm{d}\xi^j_{(a)} \; . \tag{3.66}$$

We can now start computing $L_Z(\eta)$; for this it will be convenient to define the $(4n \times 4n)$ matrix $F_j{}^i = \partial_j f^i = (\partial f^i / \partial x^j)$. In fact, we have

$$Z \lrcorner \omega = f^i \, A_{ij} \, \mathrm{d}x^j \; ;$$
$$\mathrm{d}(Z \lrcorner \omega) = (\partial_\ell f^i) \, A_{ij} \, \mathrm{d}x^\ell \wedge \mathrm{d}x^j \; = \; F_\ell{}^i \, A_{ij} \, \mathrm{d}x^\ell \wedge \mathrm{d}x^j \; . \tag{3.67}$$

With the "block decomposed" notation, we have $Z \lrcorner \omega^{(a)} = f^{4(\kappa-1)+i} A^{(\kappa)}_{ij} \mathrm{d}\xi^j_{(a)}$, and therefore (with a maybe not so nice "mixed" notation)

$$\mathrm{d}(Z \lrcorner \omega^{(a)}) \; = \; \frac{\partial f^{4(a-1)+i}}{\partial x^\ell} \, A^{(a)}_{ij} \, \mathrm{d}x^\ell \wedge \mathrm{d}\xi^j_{(a)} \; = \; F_\ell{}^{4(a-1)+i} \, A^{(a)}_{ij} \, \mathrm{d}x^\ell \wedge \mathrm{d}\xi^j_{(a)} \; .$$

Finally, we have $L_Z(\eta) = F_\ell{}^i A_{ij} A_{km} \mathrm{d}x^\ell \wedge \mathrm{d}x^j \wedge \mathrm{d}x^k \wedge \mathrm{d}x^m$; with the block decomposed notation, this reads

$$L_Z(\eta) \; = \; \sum_a \sum_b F_\ell{}^{4(a-1)+i} \, A^{(a)}_{ij} \, A^{(b)}_{km} \mathrm{d}x^\ell \wedge \mathrm{d}\xi^j_{(a)} \wedge \mathrm{d}\xi^k_{(b)} \wedge \mathrm{d}\xi^m_{(b)} \; . \tag{3.68}$$

We have to apply ι^*_κ on this, or better on each of the terms in the sum. The result will be nonzero *only* if *both* a and b are equal to κ; moreover, only derivatives w.r.t. variables in the κ-th block will contribute. That is, we have

$$\iota^*_\kappa[L_Z(\eta)] = \iota^*_\kappa \left[\sum_a \sum_b F_\ell{}^{4(a-1)+i} \, A^{(a)}_{ij} \, A^{(b)}_{km} \mathrm{d}x^\ell \wedge \mathrm{d}\xi^j_{(a)} \wedge \mathrm{d}\xi^k_{(b)} \wedge \mathrm{d}\xi^m_{(b)} \right]$$

$$= \sum_a \sum_b \iota^*_\kappa \left[F_\ell{}^{4(a-1)+i} \, A^{(a)}_{ij} \, A^{(b)}_{km} \mathrm{d}x^\ell \wedge \mathrm{d}\xi^j_{(a)} \wedge \mathrm{d}\xi^k_{(b)} \wedge \mathrm{d}\xi^m_{(b)} \right]$$

$$= \iota^*_\kappa \left[F_\ell{}^{4(\kappa-1)+i} \, A^{(\kappa)}_{ij} \, A^{(\kappa)}_{km} \mathrm{d}x^\ell \wedge \mathrm{d}\xi^j_{(\kappa)} \wedge \mathrm{d}\xi^k_{(\kappa)} \wedge \mathrm{d}\xi^m_{(\kappa)} \right]$$

$$= \left[\iota_\kappa^*(F_\ell^{\,4(\kappa-1)+i}\mathrm{d}x^\ell) \right] \wedge \left(A_{ij}^{(\kappa)}\, A_{km}^{(\kappa)}\mathrm{d}\xi_{(\kappa)}^j \wedge \mathrm{d}\xi_{(\kappa)}^k \wedge \mathrm{d}\xi_{(\kappa)}^m \right)$$

$$= \sum_{s=1}^{4} \frac{\partial f_\ell^{\,4(\kappa-1)+i}}{\partial \xi_{(\kappa)}^s} \left(A_{ij}^{(\kappa)}\, A_{km}^{(\kappa)} \right) \mathrm{d}\xi_{(\kappa)}^s \wedge \mathrm{d}\xi_{(\kappa)}^j \wedge \mathrm{d}\xi_{(\kappa)}^k \wedge \mathrm{d}\xi_{(\kappa)}^m$$

Denoting by $F^{(\kappa)}$ the restriction of F to the κ-th block (that is, decomposing F in (4×4) blocks, this is the κ-th block on the diagonal), we have in the end

$$L_Z(\eta) \;=\; \sum_{\kappa=1}^{n} \left[(F^{(\kappa)})_i^{\,\ell}\, A_{\ell j}^{(\kappa)}\, A_{pq}^{(\kappa)} \right] \mathrm{d}\xi_{(\kappa)}^i \wedge \mathrm{d}\xi_{(\kappa)}^j \wedge \mathrm{d}\xi_{(\kappa)}^p \wedge \mathrm{d}\xi_{(\kappa)}^q . \tag{3.69}$$

Now we just have to note that in (3.69) the indices $\{i, j, p, q\}$ must all be different (or we get a zero contribution) and are in the range $1, \ldots, 4$; if we think of p and q ($p \neq q$) as assigned, then i and j only have two possible values, and it must be $i \neq j$. On the other hand, A is skew-symmetric, so if $\ell = j$ we get no contribution. The result of all this is that only terms with $i = \ell$ enter in the sum.[7]
In view of this remark, Eq. (3.69) reads also

$$L_Z(\eta) \;=\; \sum_{\kappa=1}^{n} \mathrm{div}(f^{(\kappa)}) \left[A_{ij}^{(\kappa)}\, A_{pq}^{(\kappa)}\, \mathrm{d}\xi_{(\kappa)}^i \wedge \mathrm{d}\xi_{(\kappa)}^j \wedge \mathrm{d}\xi_{(\kappa)}^p \wedge \mathrm{d}\xi_{(\kappa)}^q \right] . \tag{3.70}$$

The term in square brackets corresponds to $8\mathcal{P}_2(A^{(\kappa)})$, where $\mathcal{P}_2$ is the invariant introduced before. It was shown in [68] (see also (3.3) in Remark 3.1 above) that, in particular, $\mathcal{P}_2(A) = \pm 1$ for four dimensional matrices A corresponding to positively or negatively oriented standard hyperkahler structures. This means that, according to (3.70), we can have $L_Z(\eta) = 0$ if and only if

$$\mathrm{div}(f^{(\kappa)}) \;=\; 0 \quad \forall \kappa = 1, \ldots, n . \tag{3.71}$$

Recalling now the discussion of Sect. 3.3.1, we conclude that a vector field is canonical if and only if its restriction to any of the invariant $\mathbf{R}^4$ subspaces is a Dirac vector field for the HK structure restricted to the subspace. (This should also be compared with Lemma 3.6, and Remark 2.1 in Chap. 2).

In view of Definition 2.2 in Chap. 2, and recalling that here we are working under the assumption (3.65), this shows that the vector fields Z in $\mathbf{R}^{4n}$ such that $L_Z[\iota_{(\alpha)}^*(\omega \wedge \omega)] = 0$ (for all $\alpha = 1, \ldots, n$) are necessarily Dirac.

In other words, (3.65) and (3.71) show that the problem decomposes into the four-dimensional problems relative to each invariant $\mathbf{R}^4$ subspace, and on these the discussion of Sect. 3.3.1 applies. $\triangle$

---

[7]One can also easily checked this is true by explicit computations, considering a generic symmetric matrix F and A a generic linear combination of the Y_α^0 (or of the $\widehat{Y}_\alpha^0$).

Remark 3.9 We have thus shown that canonical vector fields and Dirac ones do coincide (at least in the Euclidean case under study here). It is obvious that the canonical vector fields do form a Lie algebra, as they are characterized by preservation of a structure. It immediately follows that the Dirac vector fields (w.r.t. a given Euclidean hyperkahler structure) do also form a Lie algebra.   ⊙

Remark 3.10 The $\mathcal{P}$-operator introduced in Sect. 3.1 (see Remark 3.1; see also [68]) can also be used to detect canonical vector fields (i.e., in the light of Lemma 3.6, Dirac vector fields). In fact, for a generic vector field $Z = f^i \partial_i$ and a symplectic structure $\omega = (1/2) K_{ij} \mathrm{d}x^i \wedge \mathrm{d}x^j$, we have (3.67) (with K taking the place of A); it follows that

$$\mathcal{L}_Z[(1/2)\,\omega \wedge \omega] = (\partial_m f^\ell)\,K_{\ell s}\,K_{ij}\,\mathrm{d}x^m \wedge \mathrm{d}x^s \wedge \mathrm{d}x^i \wedge \mathrm{d}x^j \ .$$

This shows at once that $\mathcal{L}_Z(\omega \wedge \omega) = 0$ (i.e. Z is canonical) if and only if

$$\epsilon^{msij}\,(\partial_m f^\ell)\,K_{\ell s}\,K_{ij} = 0 \ . \tag{3.72}$$

But the left hand side of this equation is immediately rewritten in terms of the operator $\mathcal{P}_1$ as

$$\mathcal{P}_1[FK, K] := \epsilon^{ijk\ell}\,(\partial_i f^m)\,K_{mj}\,K_{k\ell} \ .$$

Thus we conclude that the vector field Z is canonical for a hyperkahler structure if and only if the matrix F of components $F_i{}^m = \partial_i f^m$ built out of its components satisfies

$$\mathcal{P}_1[FK, K] = 0 \tag{3.73}$$

for all the K associated to all the $\omega = c_\alpha \omega_\alpha$ (here ω_α are the symplectic forms generating the hypersymplectic structure); see also the discussion in [68] for details. ⊙

3.4 Dirac versus Hyperhamiltonian Canonical Vector Fields

We have shown in Sect. 3.3.2 that any one-parameter smooth family of canonical transformations corresponds to the flow of a Dirac field. It is natural to wonder if one can identify, among these, those which correspond to a hyperhamiltonian vector field.

Our discussion, see in particular Sect. 3.3, and the very definition of Dirac vector fields, show that the problem decomposes into a problem in each four-dimensional $\mathbf{R}_{(k)}^4$ block. We can thus just discuss the problem in $M = \mathbf{R}^4$.

The explicit expression of the f^i in terms of the $\mathcal{H}^a$ and $\widehat{\mathcal{H}}^a$ can also be used to devise a test to ascertain if a given vector field $Z = f^i \partial_i$ in $\mathbf{R}^4$ is hyperhamiltonian w.r.t. a standard hyperkahler structure; note that a preliminary condition is that it should have zero divergence (this only shows that the vector field is a Dirac one). It is convenient to use the representation (3.50) for Ψ.

Lemma 3.7 *Let $M = \mathbf{R}^4$ and Z be a vector field in M; let $\{\Psi_Z\} \subset \Lambda^2(M)$ be the set of forms associated to Z as in the proof to Lemma 3.6. Then Z is hyperhamiltonian w.r.t. the standard positively oriented hyperkahler structure if and only if there exists $\Psi = (1/2)P_{ij}dx^i \wedge dx^j \in \{\Psi_Z\}$ such that its components satisfy the equations*

$$P_{23} = P_{14}, \quad P_{24} = -P_{13}, \quad P_{34} = P_{12}; \tag{3.74}$$

in this case the Hamiltonians are identified by the components of Ψ as

$$\mathcal{H}^1 = P_{12}, \quad \mathcal{H}^2 = P_{13}, \quad \mathcal{H}^3 = P_{14}. \tag{3.75}$$

Similarly, X is hyperhamiltonian w.r.t. the standard negatively oriented hyperkahler structure if and only if the components of Ψ satisfy the equations

$$P_{23} = -P_{14}, \quad P_{24} = P_{13}, \quad P_{34} = -P_{12}; \tag{3.76}$$

in this case the Hamiltonians are identified by the components of Ψ as

$$\widehat{\mathcal{H}}^1 = -P_{13}, \quad \widehat{\mathcal{H}}^2 = P_{14}, \quad \widehat{\mathcal{H}}^3 = P_{12}. \tag{3.77}$$

Proof This follows from (3.53) and (3.54) with standard computations. △

Remark 3.11 The form Ψ corresponds, see (3.50), to a skew-symmetric matrix P; our Lemma 3.7 shows that the vector fields which are hyperhamiltonian w.r.t. the positively or negatively oriented hyperkahler structure—and which we will call, with an abuse of language, "positive" and "negative" respectively—are characterized by two-forms Ψ which correspond to matrices $P_{(\pm)}$ satisfying

$$P_{23} = \pm P_{14}, \quad P_{24} = \mp P_{13}, \quad P_{34} = \pm P_{12}. \tag{3.78}$$

In other words, the $P_{(\pm)}$ are given by

$$P_{(\pm)} = \begin{pmatrix} 0 & \eta_1 & \eta_2 & \eta_3 \\ -\eta_1 & 0 & \pm\eta_3 & \mp\eta_2 \\ -\eta_2 & \mp\eta_3 & 0 & \pm\eta_1 \\ -\eta_3 & \pm\eta_2 & \mp\eta_1 & 0 \end{pmatrix}. \tag{3.79}$$

We could of course also use (3.74), (3.76)—or equivalently (3.53) and (3.54)—and write these in terms of the Hamiltonians; we omit explicit formulas. ☉

Remark 3.12 We have defined in Remark 1.3 of Chap. 1, Sect. 1.4.3, the operator $\mathcal{P}_m$ acting on $(2m \times 2m)$ antisymmetric matrices; see also Remarks 3.1 and 3.10. In this case, we deal with four dimensional matrices and hence $\mathcal{P}_2$. This operator is able to detect the orientation of symplectic matrices, and representing in a certain sense the square root of their determinant (see [68] for details). It is then immediately checked, by explicit computation, that

$$\mathcal{P}_2 \left[P_{(\pm)} \right] = \pm \left(\eta_1^2 + \eta_2^2 + \eta_3^2 \right) . \tag{3.80}$$

In other words, this operator—or more precisely the sign of the result of applying it to P—also detects the "orientation" (in the sense specified above) of a hyperhamiltonian vector field by applying it on the (matrix identifying the) associated form Ψ. See also the discussion in Sect. 3.5.   ⊙

Remark 3.13 Using the explicit expression for the M_a, we obtain that in the positively oriented case Ψ is simply $\Psi = \Psi_{(+)} = \mathcal{H}^1 \omega_1 + \mathcal{H}^2 \omega_2 + \mathcal{H}^3 \omega_3$. Similarly, in the negatively oriented case we have $\Psi = \Psi_{(-)} = -(\widehat{\mathcal{H}}^1 \widehat{\omega}_1 + \widehat{\mathcal{H}}^2 \widehat{\omega}_2 + \widehat{\mathcal{H}}^3 \widehat{\omega}_3)$. In the general case (that is, $X = Y + \widehat{Y}$) we obtain just the sum $\Psi = \Psi_{(+)} + \Psi_{(-)}$ of the forms corresponding to the positively and the negatively oriented parts of the vector field.   ⊙

Remark 3.14 It follows from the formulas (3.52) above that if we *know* that a vector field is hyperhamiltonian w.r.t. the positively oriented standard hyperkahler structure, then we can have information on the Hamiltonians by computing e.g.

$$\partial_2 f^1 - \partial_1 f^2 + \partial_4 f^3 - \partial_3 f^4 = \Delta \mathcal{H}^1 ,$$
$$\partial_3 f^1 - \partial_4 f^2 - \partial_1 f^3 + \partial_2 f^4 = \Delta \mathcal{H}^2 ,$$
$$\partial_4 f^1 + \partial_3 f^2 - \partial_2 f^3 - \partial_1 f^4 = \Delta \mathcal{H}^3 ;$$

and similar formulas if we *know* that a vector field is hyperhamiltonian w.r.t. the negatively oriented standard hyperkahler structure.   ⊙

3.5 Dirac Vector Fields and Self-duality

We will now briefly discuss the relation between standard hyperkahler structures in $\mathbf{R}^4$ and self-dual (and anti-self-dual) two-forms.

As mentioned above, while the form $\Phi = Z \lrcorner \Omega$ is uniquely defined by the vector field Φ, the form Ψ such that $\Phi = d\Psi$ is not uniquely defined; more precisely, it is defined up to a closed form $\delta \Psi$. As we are in $\mathbf{R}^{4n}$, closed forms are also exact, i.e. necessarily there exists a form Γ such that $\delta \Psi = d\Gamma$.

As our discussion can be reduced to the four dimensional case, we should consider this situation is such a setting[8]; it turns out this presents some peculiar—and interesting—features.

These are related to the action of the Hodge star operator $\star$ on $\Lambda(\mathbf{R}^4)$. We recall that in general $\star : \Lambda^k(\mathbf{R}^q) \to \Lambda^{q-k}(\mathbf{R}^q)$, and for any $\beta \in \Lambda^k(\mathbf{R}^q)$, we have $\star(\star\beta) = (-1)^{k(q-k)}\beta$.

In the case $q = 4, k = 2$ we have $\star : \Lambda^2(\mathbf{R}^4) \to \Lambda^2(\mathbf{R}^4)$; for $\beta \in \Lambda^2(\mathbf{R}^4)$ we just have $\star(\star\beta) = \beta$.

It follows that there are $\beta \in \Lambda^2(\mathbf{R}^4)$ such that $\star\beta = \pm\beta$; these forms β are accordingly called *self-dual* or *anti-self-dual* and they span $\Lambda^2(\mathbf{R}^4)$. In other words, as well known,

$$\Lambda^2(\mathbf{R}^4) = \Lambda_+^2(\mathbf{R}^4) \oplus \Lambda_-^2(\mathbf{R}^4),$$

where Λ_+^2 (Λ_-^2) is the space of self-dual (of anti-self-dual) forms.

Remark 3.15 Denoting by ζ the one-form dual (in the sense of the Riemannian metric, in this case the Euclidean one) to the vector field Z, it is easy to check that our definition of Φ corresponds to $\Phi = \star\zeta$. Thus $\Psi \in \Lambda^2(\mathbf{R}^4)$ is a form satisfying $\mathrm{d}\Psi = \star\zeta$. ⊙

Lemma 3.8 *Let* $\Psi = (1/2)P_{ij}\mathrm{d}x^i \wedge \mathrm{d}x^j \in \Lambda^2(\mathbf{R}^4)$. *Then* Ψ *is self-dual (anti-self-dual) if and only if its components satisfy*

$$P_{24} = -s\,P_{13}, \quad P_{14} = s\,P_{23}, \quad P_{12} = s\,P_{34},$$

with $s = +1$ ($s = -1$).

Proof This is just a consequence of the definition of the $\star$ operator. In fact, in components we have

$$\begin{aligned}
\Psi = {} & P_{12}\mathrm{d}x^1 \wedge \mathrm{d}x^2 + P_{13}\mathrm{d}x^1 \wedge \mathrm{d}x^3 + P_{14}\mathrm{d}x^1 \wedge \mathrm{d}x^4 \\
& + P_{23}\mathrm{d}x^2 \wedge \mathrm{d}x^3 + P_{24}\mathrm{d}x^2 \wedge \mathrm{d}x^4 + P_{34}\mathrm{d}x^3 \wedge \mathrm{d}x^4 \\
\star\Psi = {} & P_{12}\mathrm{d}x^3 \wedge \mathrm{d}x^4 + P_{13}\mathrm{d}x^4 \wedge \mathrm{d}x^2 + P_{14}\mathrm{d}x^3 \wedge \mathrm{d}x^2 \\
& + P_{23}\mathrm{d}x^4 \wedge \mathrm{d}x^1 + P_{24}\mathrm{d}x^3 \wedge \mathrm{d}x^1 + P_{34}\mathrm{d}x^1 \wedge \mathrm{d}x^2
\end{aligned}$$

Then the statement follows immediately. △

Lemma 3.9 *Let* Z *be a divergence-free vector field in* $\mathbf{R}^4$ *equipped with the Euclidean metric, and* ζ *the one-form dual to it in the sense of the Riemannian metric. Let* $\{\Psi\} \in \Lambda^2(\mathbf{R}^4)$ *be the set of forms satisfying* $\mathrm{d}\Psi = \star\zeta$. *The set* $\{\Psi\}$ *contains a self-dual (anti-self-dual) form* Ψ *if and only if* Z *is a hyperhamiltonian vector field with respect to a positively (negatively) oriented standard hyperkähler structure.*

---

[8]In this case, rather trivially, $\Phi \in \Lambda^3(\mathbf{R}^4)$, and $\Psi \in \Lambda^2(\mathbf{R}^4)$. Moreover, $\Gamma \in \Lambda^1(\mathbf{R}^4)$.

Proof The vector field Z is divergence-free, and hence Dirac, if there exist two triplets of Hamiltonians, $\mathcal{H}_i$, $\widehat{\mathcal{H}}_i$ satisfying (3.52) Using the explicit expression of the matrices M_a and $\widehat{M}_a$, we get (3.52), and hence (3.53) and (3.54). Then, Ψ is self-dual if and only if $\widehat{\mathcal{H}}_a = 0$; and correspondingly it is anti-self-dual if and only if $\mathcal{H}_a = 0$.   △

Remark 3.16 The operator $\mathcal{P}_2$ considered above is therefore able to detect self-duality (anti-self-duality) of forms in $\Lambda^2(\mathbf{R}^4)$.   ⊙

Chapter 4
Hyperkahler Maps

4.1 Hyperkahler Transformations

In this chapter, we will describe the (connected component of the) group of hyperkahler transformations (**Q**-maps) for Euclidean spaces $\mathbf{R}^{4n}$ of arbitrary dimension $4n$. In this particular case one is able to provide a fairly complete characterization of the Lie algebra of this group, the *invariance algebra* L_n below, in arbitrary dimension.

We show by a completely explicit procedure (based on standard linear algebra plus some general results from the theory of Lie algebras), that

$$L_n = \operatorname{su}(2) \oplus \operatorname{sp}(n) = \operatorname{sp}(1) \oplus \operatorname{sp}(n) . \tag{4.1}$$

We also show that the "strong invariance algebra", leaving each of the ω_α invariant, is $\mathcal{G}_n = \operatorname{sp}(n)$.

These results have been known in the differential geometric literature [100, 104, 133]; but they were obtained in a rather abstract way, while the derivation we provide here is fully explicit.

These results are of course also in agreement with Berger's list of holonomy groups for Riemannian manifolds [21] and further research on this topic (see [100, 104, 133] for a comprehensive exposition of this subject; see also [74, 115], and Appendix A).

Regarding the notation for symplectic groups, $\operatorname{Sp}(n)$ will be the set of $(2n \times 2n)$ (complex) unitary symplectic matrices (thus with real representation of dimension $4n$), with Lie algebra $sp(n) \subset \operatorname{Mat}(2n, \mathbf{C}) \simeq \operatorname{Mat}(4n, \mathbf{R})$.

We would like to stress the *physical* relevance of the flat case. This is due not only to the fact the Dirac equation is set in flat Minkowski space (which would maybe suffice by itself), but also to the fact that most of the physically relevant non-trivial hyperkahler manifolds are obtained from higher dimensional Euclidean $\mathbf{R}^{4n}$ manifolds (with standard hyperkahler structure) via the momentum map construction pioneered by Hitchin et al. [92]. Thus their hyperkahler structures are obtained,

© Springer International Publishing AG 2017
G. Gaeta and M.A. Rodríguez, *Lectures on Hyperhamiltonian Dynamics and Physical Applications*, Mathematical Physics Studies, DOI 10.1007/978-3-319-54358-1_4

through the same construction, from those of the higher dimensional Euclidean manifold; e.g. the hyperkahler structure (beside of course the metric) for the Taub-NUT manifold [49, 126, 141] can be built explicitly starting from those in $\mathbf{R}^8$ [67]. Thus, albeit maybe not so interesting for Geometry, the Euclidean case has a substantial relevance for Physics, and we believe it is worth to provide a fully explicit discussion of the invariance group for hyperkahler structures in Euclidean $\mathbf{R}^{4n}$ spaces.

We will use the notations introduced in the previous sections. Let us recall (see Sect. 1.4) that if $\mathbf{J}$ and $\widehat{\mathbf{J}}$ are different hyperkahler structures on the same Riemannian manifold $(\mathcal{V}, g)$ and if each of them can be expressed in terms of the other,

$$\widehat{J}_\alpha = \sum_{\beta=1}^{3} R_{\alpha\beta} J_\beta \,, \tag{4.2}$$

the two structures are said to be *equivalent*. An equivalence class of hyperkahler structures on $(\mathcal{V}, g)$ is said to be a **Q**-*structure* on $(\mathcal{V}, g)$.

It should be stressed that since both the J and the $\widehat{J}$ satisfy the quaternionic relations (1.15), necessarily the matrix R with entries $R_{\alpha\beta}$ in (4.2) belongs to the Lie group SO(3). Moreover, as both J and $\widehat{J}$ are covariantly constant, it follows that $\nabla R = 0$ as well.

If we define local coordinates in $\mathcal{V}$, the $(1,1)$ tensors J_α are represented by matrices (which we denote again by J_α with a standard abuse of notation), and the quaternionic relation (1.15) holds between such matrices.

4.1.1 Maps on Hyperkahler Manifolds

We would now like to characterize the maps $\varphi : \mathcal{V} \to \mathcal{V}$ which leave invariant the hyperkahler structure, or at least the equivalence class of hyperkahler structures discussed above, i.e. the **Q**-structure on $(\mathcal{V}, g)$.

If $\varphi : \mathcal{V} \to \mathcal{V}$ is an arbitrary smooth map in $\mathcal{V}$, the hyperkahler structure will change according to the rule of transformations of $(1, 1)$ tensors (which amounts to a conjugation), i.e.

$$J_\alpha \to \widetilde{J}_\alpha := \Lambda J_\alpha \Lambda^{-1} \,,$$

where Λ is the Jacobian of φ.

We recall (see Sect. 2.7) the definition of strongly hyperkahler maps: the map $\varphi : \mathcal{V} \to \mathcal{V}$ is *strongly* hyperkahler for $(\mathcal{V}, g, \mathbf{J})$ if it preserves both the metric g and the hyperkahler structure $\mathbf{J}$.

It is obvious that the set of strongly hyperkahler maps for $(\mathcal{V}, g, \mathbf{J})$ is a group. Such maps are also called *tri-holomorphic*, as they are holomorphic for each of the three complex structures J_α. The group of strongly hyperkahler maps on $(\mathcal{V}, g; \mathbf{J})$ will be called the *strong* hyperkahler *group* on $(\mathcal{V}, g; \mathbf{J})$; or for short the *strong invariance group of* $\mathcal{V}$. Correspondingly, its elements will be called, with an abuse of language, *strong invariance maps*.

We have also defined what a hyperkahler map is (Sect. 2.7): the map $\varphi : \mathcal{V} \to \mathcal{V}$ is *hyperkahler* for $(\mathcal{V}, g, \mathbf{J})$ if it preserves the metric and maps the hyperkahler structure into an equivalent one. In this case it is also said to be a **Q**-*map*, as it preserves the **Q**-structure on $(\mathcal{V}, g)$.

Here again it is obvious that that the set of hyperkahler maps for $(\mathcal{V}, g, \mathbf{J})$ is a group. This will be called the hyperkahler group on $(\mathcal{V}, g; \mathbf{J})$; or for short the *invariance group of* $\mathcal{V}$. Correspondingly, its elements will be called, with an abuse of language, *invariance maps*. Any strong invariance map is also an invariance map; the set of maps which are hyperkahler but not strongly hyperkahler will also be denoted as *regular invariance maps*.

Remark 4.1 It is clear that the maps preserving the hyperkahler structure will also preserve the hypersymplectic one, and those mapping the hyperkahler structure into an equivalent one will also maps the hypersymplectic structure into an equivalent one. Thus it would also be legitimate to denote the maps and groups identified above as strongly hypersymplectic and respectively hypersymplectic ones; the groups will correspondingly be called the *strong hypersymplectic group* and the *hypersymplectic group*.   ⊙

4.1.2 Orientation

As recalled above, a hyperkahler manifold is *oriented*; we can thus consider in particular orientation-switching maps $\mathcal{P}$, obviously satisfying $\mathcal{P}^2 = I$. Under such a map the hyperkahler structure will not be preserved; note that the Riemannian metric should instead be invariant under such an orientation-switching map (this will in particular be the case for the Euclidean metric for any orthogonal $\mathcal{P}$). From now on we will only consider maps $\mathcal{P}$ preserving the metric. Under these maps, the hyperkahler structure will be mapped to the (non-equivalent) *dual* one [68].

It is quite obvious that hyperkahler structures which are dual to each other (we refer to these as a *dual pair*) are strongly related; it also turns out that in some physical applications of hyperhamiltonian dynamics (in particular, in the description of the Dirac equation in hyperhamiltonian terms [66]) one needs both elements of a dual pair.

When we work in the symplectic framework, so that the hyperkahler structure corresponds to a triple of symplectic structures $\{\omega_\alpha\}$, the action of the map $\mathcal{P}$ on these is simply given by the pull-back. This induces a conjugation between the ω_α and the dual ones, $\widetilde{\omega}_\alpha = \mathcal{P}^* \omega_\alpha$, and hence between a hyperkahler structure and its dual one. This shows that hyperkahler structures related by such an orientation switch are conjugated. (Representing the forms ω_α and the complex structures J_α in coordinates, the conjugation is described by the action of a matrix P which is orthogonal with respect to the metric).

We conclude that hyperkahler structures related by such a map will be invariant under isomorphic groups of transformations G and $\widehat{G}$, with $\widehat{G} = P^{-1}GP$.

In the following we will have to consider spaces $\mathbf{R}^{4n}$ and the possibility to independently switch orientation in the $\mathbf{R}^4$ subspaces on which (the appropriate restriction of) the $\omega_\alpha \wedge \omega_\alpha$ give a volume form; the same considerations presented above will also hold for the restriction of the hyperkahler structure to each of these subspaces, and this will be rather useful to simplify our computations.

4.2 Euclidean Spaces

We will specialize our general notions and discussions to the Euclidean case.

4.2.1 Hyperkahler Structures

As discussed in Sect. 1.4.3, the simplest example of a hyperkahler manifold is $\mathbf{R}^{4n}$ with the Euclidean metric, equipped with the *standard hyperkahler structures* which we detail again below (since they will play a crucial role in our discussion). It should be noted that, since the metric is here Euclidean, the covariant derivative is the usual derivative (the Levi-Civita connection is trivial). Then $\partial_{x_i} J(x) = 0$ for $i = 1, \ldots, 4n$, and the hyperkahler structure is actually constant.

The standard positively and negatively oriented standard hyperkahler structures in $\mathbf{R}^4$ are given respectively by

$$Y_1 = \begin{pmatrix} 0 & 1 & 0 & 0 \\ -1 & 0 & 0 & 0 \\ 0 & 0 & 0 & 1 \\ 0 & 0 & -1 & 0 \end{pmatrix}, \quad Y_2 = \begin{pmatrix} 0 & 0 & 0 & 1 \\ 0 & 0 & 1 & 0 \\ 0 & -1 & 0 & 0 \\ -1 & 0 & 0 & 0 \end{pmatrix}, \quad Y_3 = \begin{pmatrix} 0 & 0 & 1 & 0 \\ 0 & 0 & 0 & -1 \\ -1 & 0 & 0 & 0 \\ 0 & 1 & 0 & 0 \end{pmatrix}, \quad (4.3)$$

$$\widehat{Y}_1 = \begin{pmatrix} 0 & 0 & 1 & 0 \\ 0 & 0 & 0 & 1 \\ -1 & 0 & 0 & 0 \\ 0 & -1 & 0 & 0 \end{pmatrix}, \quad \widehat{Y}_2 = \begin{pmatrix} 0 & 0 & 0 & -1 \\ 0 & 0 & 1 & 0 \\ 0 & -1 & 0 & 0 \\ 1 & 0 & 0 & 0 \end{pmatrix}, \quad \widehat{Y}_3 = \begin{pmatrix} 0 & -1 & 0 & 0 \\ 1 & 0 & 0 & 0 \\ 0 & 0 & 0 & 1 \\ 0 & 0 & -1 & 0 \end{pmatrix}. \quad (4.4)$$

In the four-dimensional case ($n = 1$), it is easily checked that any constant hyperkahler structure J_α, i.e. any set of constant skew-symmetric matrices satisfying the quaternionic relations (1.15), can be transformed through a conjugation with a matrix $P \in SO(4)$ into one of the two inequivalent (under $SO(4)$ conjugation) sets of matrices Y_α and $\widehat{Y}_\alpha$, $\alpha = 1, 2, 3$. (Note that indeed any skew-symmetric matrix in $\mathbf{R}^4$ is written as a sum of the Y_α and of the $\widehat{Y}_\alpha$.) This corresponds to the su(2) algebra having two irreducible representations in $\mathbf{R}^4$.

A similar result holds in $\mathbf{R}^{4n}$: in this case one acts with $G = SO(4n)$, and the hyperkahler structures can be transformed into some direct sum of the above standard ones; we stress in this sum there will in general be blocks of each orientation. More precisely we have the following

Lemma 4.1 *Given any hyperkahler structure* $\{J_\alpha\}$ *in* $\mathbf{R}^{4n}$, *there exists a conjugation given by a regular matrix* $P \in SO(4n)$, *such that* $\widetilde{J}_\alpha := P J_\alpha P^{-1}$ *are diagonal* 4×4 *block matrices, and the blocks in the diagonal are equal to either* Y_α *or* $\widehat{Y}_\alpha$.

Proof As we are in Euclidean spaces, $\nabla J_\alpha = 0$ means the J_α are actually constant; thus they provide a (real, quaternionic) representation of $SU(2)$. This representation can be decomposed as the sum of real quaternionic irreducible representations, which are well known (see e.g. chap. 8 of [105]) to be four dimensional.   △

4.2.2 Hyperkahler and Strongly Hyperkahler Maps

Let us now consider (strongly) hyperkahler maps in Euclidean spaces; in this case we can be quite more specific, due to the specially simple metric and the triviality of the Levi-Civita connection.

If g is the matrix associated to the metric and Λ is the Jacobian of a transformation φ in $\mathcal{V}$, the change in the metric (or more precisely in the coordinate representation of the metric) is

$$g \rightarrow \widetilde{g} = \Lambda g \Lambda^T ; \tag{4.5}$$

for the Euclidean metric, $g = I_4$ and hence $\widetilde{g} = \Lambda \Lambda^T$; thus $\widetilde{g} = g$ requires

$$\Lambda \Lambda^T \equiv I_4 \tag{4.6}$$

and it should be $\Lambda(x) \in O(4n)$ (or $\Lambda(x) \in SO(4n)$ if we want to keep the orientation) for any $x \in \mathcal{V}$.

Moreover, as both the original and the transformed complex structures should be covariantly constant, it should also be $\nabla \Lambda = 0$: but since here ∇ is the trivial connection, this means $\partial_i \Lambda(x) = 0$ for all $i = 1, ..., 4n$, hence Λ is constant.

This shows at once that hyperkahler and strongly hyperkahler maps should be constant orthogonal (or special orthogonal if we want to preserve orientation) ones, in full generality.

4.2.2.1 Strongly Hyperkahler Maps

Let us now consider in detail strongly hyperkahler maps, and represent the J_α by means of the corresponding matrices in coordinates. Regarding the preservation of the hyperkahler structure, we should impose in this (strong invariance) case

$$\Lambda J_\alpha \Lambda^{-1} = J_\alpha ; \tag{4.7}$$

this means that Λ should commute with each of the J_α. Since these matrices are a representation of su(2), we could apply representation theory to this problem.

Indeed, let us consider a set of three $m \times m$ real matrices, $J_\alpha, \alpha = 1, 2, 3$ satisfying $J_\alpha J_\beta = \epsilon_{\alpha\beta\gamma} J_\gamma - \delta_{\alpha\beta} I_m$. This relation implies $[J_\alpha, J_\beta] = 2\epsilon_{\alpha\beta\gamma} J_\gamma$, stating that the matrices

$$\Gamma_\alpha = (1/2)J_\alpha \quad (\alpha = 1, 2, 3)$$

form a representation $\mathcal{R}$ of su(2). But it also yields $J_\alpha^2 = -I_m$, $\Gamma_\alpha^2 = -(1/4)I_m$. This property implies that the eigenvalues of the Casimir $-\sum_\alpha \Gamma_\alpha^2$ are (3/4) and then the representation is (complex) reducible into a direct sum of a certain number of $(\frac{1}{2})$ representations:

$$\mathcal{R} = \left(\frac{1}{2}\right) \oplus \cdots \oplus \left(\frac{1}{2}\right) .$$

Since we have real matrices, the representation is real reducible to a diagonal block form, with 4×4 blocks, each of them associated to a $(\frac{1}{2}) \oplus (\frac{1}{2})$ representation of su(2) with real matrices and the dimension is $m = 4n$.

4.2.2.2 Hyperkahler Maps

Let us now consider **Q**-maps, i.e. hyperkahler ones. The requirement to map **J** into a possibly different, but equivalent, structure $\widetilde{\mathbf{J}}$ implies that

$$\widetilde{J}_\alpha = \Lambda J_\alpha \Lambda^{-1} = \sum_{\beta=1}^{3} R_{\alpha\beta} J_\beta, \quad \sum_{\beta=1}^{3} R_{\alpha\beta}^2 = 1 , \quad \alpha = 1, 2, 3 . \tag{4.8}$$

In the same way as in the strong version, the first condition implies that the matrices Λ should be in O(4n), and in SO(4n) if we want to leave invariant the orientation. As for the second condition, it yields the following constraint. The new matrices $\widetilde{J}_\alpha$ should satisfy the quaternionic relations (1.15), i.e.

$$\widetilde{J}_\alpha \widetilde{J}_\beta = \epsilon_{\alpha\beta\gamma} \widetilde{J}_\gamma - \delta_{\alpha\beta} I . \tag{4.9}$$

Substituting (sum over repeated indices is assumed)

$$R_{\alpha\mu} R_{\beta\nu} J_\mu J_\nu = \epsilon_{\mu\nu\rho} R_{\alpha\mu} R_{\beta\nu} J_\rho + \delta_{\mu\nu} R_{\alpha\mu} R_{\beta\nu} I = \epsilon_{\alpha\beta\gamma} R_{\gamma\rho} J_\rho - \delta_{\alpha\beta} I , \tag{4.10}$$

we obtain

$$R_{\alpha\mu} R_{\beta\mu} = \delta_{\alpha\beta}, \quad \epsilon_{\mu\nu\rho} R_{\alpha\mu} R_{\beta\nu} = \epsilon_{\alpha\beta\gamma} R_{\gamma\rho} . \tag{4.11}$$

The first condition means that the matrix R is an element of O(3). The second one means that the vector product of its first and second column is the third one, which yields $R \in$ SO(3). Then, in the end we obtain the equation

$$\Lambda J_\alpha \Lambda^{-1} = \sum_{\beta=1}^{3} R_{\alpha\beta} J_\beta, \quad \alpha = 1, 2, 3, \quad \Lambda \in \mathrm{SO}(4n), \quad R \in \mathrm{SO}(3) . \qquad (4.12)$$

Thus our problem is to determine which $\Lambda \in$ SO($4n$) will satisfy Eq. (4.12) for a certain $R \in$ SO(3) (which is fixed by Λ).

The Eq. (4.12) will also be called the *finite invariance equation*.

●

4.2.3 The Infinitesimal Approach

It will be convenient, in particular in the high-dimensional case, to approach this problem from the infinitesimal point of view.

At first order in a small parameter ε, we have

$$\Lambda = I_{4n} + \varepsilon X, \quad X + X^T = 0, \qquad (4.13)$$

and, in terms of the same parameter,

$$R = I_3 + \varepsilon \mathcal{L}, \quad \mathcal{L} + \mathcal{L}^T = 0 . \qquad (4.14)$$

Equation (4.12), for any hyperkahler structure J_α, is then written at the infinitesimal level as

$$(I_{4n} + \varepsilon X) J_\alpha (I_{4n} - \varepsilon X) = \sum_{\beta=1}^{3} (\delta_{\alpha\beta} + \varepsilon \mathcal{L}_{\alpha\beta}) J_\beta , \quad \alpha = 1, 2, 3, \qquad (4.15)$$

with $X \in \mathcal{M}_{4n}(\mathbf{R}), \mathcal{L} \in \mathcal{M}_3; X + X^T = 0, \mathcal{L} + \mathcal{L}^T = 0$. That is, at first order in ε

$$[X, J_\alpha] = \sum_{\beta=1}^{3} \mathcal{L}_{\alpha\beta} J_\beta, \quad \alpha = 1, 2, 3. \qquad (4.16)$$

The Eq. (4.16) will also be called the *infinitesimal invariance equation*, or shortly (as we will mainly work in the infinitesimal approach) the *invariance equation*.

The main result we will show in this Chapter is the solution of these equations for Euclidean spaces, i.e. for $\mathcal{V} = \mathbf{R}^{4n}$ with the Euclidean metric $g = I_{4n}$.

Note that (4.16) should be seen as an equation for X *and* $\mathcal{L}$; on the other hand if we fix X, i.e. if we consider a given hyperkahler transformation, we can easily find $\mathcal{L}$ in terms of X.

As mentioned above, our main task is to characterize the group of invariance maps for $(\mathcal{V}, g, \mathbf{J})$ when $(\mathcal{V}, g)$ is $\mathbf{R}^{4n}$ with the Euclidean metric. In the infinitesimal approach, we will of course look for the Lie algebra of this group; this will be called the *invariance algebra* and denoted as L. More specifically, we will denote by L_n the invariance algebra for the hyperkahler structures in the Euclidean space $\mathbf{R}^{4n}$.

In order to grasp the problem and the approach to its solution, we find convenient to first consider the simplest (and somehow degenerate) case $n = 1$ and then the first non-degenerate case $n = 2$, before tackling the general case.

In the following we will systematically use standard cartesian coordinates in the manifold $\mathbf{R}^{4n}$, and represent the tensors J_α by real $4n$-dimensional matrices in the chosen coordinate system without further notice.

4.3 The Space $\mathbf{R}^4$

When the manifold is $\mathbf{R}^4$, the explicit characterization of hyperkahler maps can be obtained in a simple way via either the infinitesimal approach sketched above (or directly working at the finite level, see [70]). The arguments used in the discussion and the proof below are well known, but keeping them in mind will help in the study of higher dimensional cases.

4.3.1 The Infinitesimal Approach

Any skew symmetric 4×4 matrix X is necessarily a linear combination of the two sets Y_α and $\widehat{Y}_\alpha$, $\alpha = 1, 2, 3$, given above; we recall these satisfy $[Y_\alpha, \widehat{Y}_\beta] = 0$. Thus we have

$$X = \frac{1}{2} \sum_{\beta=1}^{3} c_\beta Y_\beta + \frac{1}{2} \sum_{\beta=1}^{3} \widehat{c}_\beta \widehat{Y}_\beta, \tag{4.17}$$

and the invariance equation for the positively oriented standard structure is

$$\left[\left(\frac{1}{2} \sum_{\beta=1}^{3} c_\beta Y_\beta + \frac{1}{2} \sum_{\beta=1}^{3} \widehat{c}_\beta \widehat{Y}_\beta \right), Y_\alpha \right] = \sum_{\beta=1}^{3} \mathcal{L}_{\alpha\beta} Y_\beta, \quad \alpha = 1, 2, 3. \tag{4.18}$$

Then, for $\alpha = 1, 2, 3$,

$$\frac{1}{2} \sum_{\beta=1}^{3} c_\beta [Y_\beta, Y_\alpha] = \sum_{\beta=1}^{3} \mathcal{L}_{\alpha\beta} Y_\beta, \quad \sum_{\beta=1}^{3} c_\beta \sum_{\gamma=1}^{3} \epsilon_{\beta\alpha\gamma} Y_\gamma = \sum_{\gamma=1}^{3} \mathcal{L}_{\alpha\gamma} Y_\gamma, \tag{4.19}$$

and finally,

$$\mathcal{L}_{\alpha\beta} = \sum_{\gamma=1}^{3} \epsilon_{\alpha\beta\gamma} c_{\gamma}, \quad \alpha, \beta = 1, 2, 3. \tag{4.20}$$

It follows from this that we have the

Theorem 4.1 *The invariance algebra for any hyperkahler structure in $(\mathcal{V}, g) = (\mathbf{R}^4, I_4)$ is $L_1 = so(4) \simeq su(2) \times su(2)$.*

Proof As seen above, any hyperkahler structure can be reduced to either the positively or the negatively oriented standard structure; so it suffices to consider these. We will consider the positively oriented structure; the discussion for the negatively oriented one is exactly the same, interchanging the role of the Y_α and of the $\widehat{Y}_\alpha$.

As we have noted above, the effective group, rotating Y_α, is a SO(3) subgroup, with a Lie algebra generated by the matrices Y_α. The other SO(3) subgroup, which is generated by the matrices $\widehat{Y}_\alpha$, leaves the matrices J_α invariant.

The result can be understood in terms of pure group or Lie algebra theory. In fact, the group SO(4) is not a simple group but the direct product of two SO(3) groups. Its Lie algebra has a real representation given by 4×4 matrices which splits into the direct sum of two su(2) algebras. Since they commute, the action of the whole algebra through the adjoint action is reduced to the action of one of the subalgebras on itself. This is the reason why $\mathcal{L}$, see Eq. (4.20), is in fact in the 3-dimensional representation of so(3) $\simeq$ su(2), the action of the other algebra being trivial. △

4.4 The Space R^8

In the previous Section we discussed the case $n = 1$, which is somewhat degenerate in that the hyperkahler structures in standard form consist of a single block. In this section we will tackle the first non-degenerate case, i.e. $n = 2$ or $\mathbf{R}^8$; this will present the difficulties met in the general one $\mathbf{R}^{4n}$, but for it the identification of the invariance algebra L is still rather straightforward.

First of all, we note that albeit Lemma 4.1 would lead us to deal with four different types of hyperkahler structures, the invariance groups (and algebras) for them are isomorphic; this follows from different classes of standard hyperkahler structures being conjugated by the action of matrices in O(8).

Lemma 4.2 *All hyperkahler structures in Euclidean $\mathbf{R}^8$ are conjugated under O(8).*

Proof Using Lemma 4.1, the hyperkahler structure J_α (which we recall is necessarily constant) in $\mathbf{R}^8$ endowed with the Euclidean metric, can be reduced to one of the following types (in the third case the order of the blocks can be reversed):

$$Y_\alpha^{(1)} = \begin{pmatrix} Y_\alpha & \\ & Y_\alpha \end{pmatrix}, \quad Y_\alpha^{(2)} = \begin{pmatrix} \widehat{Y}_\alpha & \\ & \widehat{Y}_\alpha \end{pmatrix}, \quad Y_\alpha^{(3)} = \begin{pmatrix} Y_\alpha & \\ & \widehat{Y}_\alpha \end{pmatrix}. \tag{4.21}$$

In fact there exist a four dimensional matrix Q which satisfies $QQ^T = \lambda I_4$ and can be chosen in $O(4) \backslash SO(4)$, i.e. in the elements of $O(4)$ with determinant equal to -1, such that $Y_\alpha = Q^{-1} \widehat{Y}_\alpha Q$ for $\alpha = 1, 2, 3$; here $Q^{-1} = Q^T$, $\det Q = -1$.

Then, if $Q_2 = \text{diag}(Q, Q) \in O(8)$ and $Q_3 = \text{diag}(I_4, Q) \in O(8)$, we get

$$Q_2^{-1} Y_\alpha^{(2)} Q_2 = Y_\alpha^{(1)}, \quad Q_3^{-1} Y_\alpha^{(3)} Q_3 = Y_\alpha^{(1)}. \tag{4.22}$$

This shows that all the standard **Q**-structures—and hence, in view of Lemma 4.1, all the **Q**-structures—in $\mathbf{R}^8$ are conjugated, as claimed in the statement. $\qquad \triangle$

Theorem 4.2 *For any hyperkahler structure in* $(V, g) = (\mathbf{R}^8, I_8)$, *the invariance algebra is*

$$L_2 = \text{su}(2) \oplus \text{sp}(2) \,.$$

Proof In view of Lemma 4.2, we can just deal with $Y_\alpha^{(1)}$. An orthogonal transformation leaving invariant the Euclidean metric I_8, is an element of $O(8)$ satisfying

$$\Lambda \Lambda^T = I_8, \quad \det \Lambda = 1; \tag{4.23}$$

at the infinitesimal level the invariance of the quaternionic relations requires, as above,

$$[X, J_\alpha] = \sum_{\beta=1}^{3} \mathcal{L}_{\alpha\beta} J_\beta, \quad \alpha = 1, 2, 3, \tag{4.24}$$

where $X = -X^T \in \mathcal{M}_8$ and $\mathcal{L} = -\mathcal{L}^T \in \mathcal{M}_3$; note that here $\Lambda \in O(8)$ implies actually $X \in \text{so}(8)$.

As in the previous case, the three matrices J_α generate an su(2) algebra which is contained in the algebra so(8) (of dimension 28) of SO(8). However, in this case, the situation is not so simple, because the whole so(8) cannot be generated by the quaternionic matrices (even considering both orientations and their combinations in the 8×8 matrices). A basis of so(8) is:

$$\begin{pmatrix} Y_\alpha & 0 \\ 0 & 0 \end{pmatrix}, \begin{pmatrix} 0 & 0 \\ 0 & Y_\alpha \end{pmatrix}, \begin{pmatrix} \widehat{Y}_\alpha & 0 \\ 0 & 0 \end{pmatrix}, \begin{pmatrix} 0 & 0 \\ 0 & \widehat{Y}_\alpha \end{pmatrix},$$

$$\begin{pmatrix} 0 & Y_\alpha \\ Y_\alpha & 0 \end{pmatrix}, \begin{pmatrix} 0 & \widehat{Y}_\alpha \\ \widehat{Y}_\alpha & 0 \end{pmatrix}, \begin{pmatrix} 0 & S_i \\ -S_i & 0 \end{pmatrix}. \tag{4.25}$$

with $\alpha = 1, 2, 3$, and S_i, $i = 1, \ldots, 10$, the set of 4×4 elementary symmetric matrices (that is, E_{jj} and $E_{jk} + E_{kj}$, where E_{jk} is the elementary matrix with 1 in the position jk and 0 elsewhere).

Let us first consider the structure $Y_\alpha^{(1)}$ and the Eq. (4.16). If we write

$$X = \begin{pmatrix} A & B \\ -B^T & C \end{pmatrix}, \quad A + A^T = 0, \quad C + C^T = 0 \tag{4.26}$$

we get

$$\left[\begin{pmatrix} A & B \\ -B^T & C \end{pmatrix}, \begin{pmatrix} Y_\alpha & 0 \\ 0 & Y_\alpha \end{pmatrix}\right] = \sum_{\beta=1}^{3} \mathcal{L}_{\alpha\beta} \begin{pmatrix} Y_\beta & 0 \\ 0 & Y_\beta \end{pmatrix} \tag{4.27}$$

and the relations

$$[A, Y_\alpha] = \sum_{\beta=1}^{3} \mathcal{L}_{\alpha\beta} Y_\beta, \quad [C, Y_\alpha] = \sum_{\beta=1}^{3} \mathcal{L}_{\alpha\beta} Y_\beta, \quad [B, Y_\alpha] = 0. \tag{4.28}$$

Using the previous results in dimension four, we obtain from the first relation

$$A = \frac{1}{2} \sum_{\beta=1}^{3} a_\beta Y_\beta + \frac{1}{2} \sum_{\beta=1}^{3} \widehat{a}_\beta \widehat{Y}_\beta,$$

$$C = \frac{1}{2} \sum_{\beta=1}^{3} c_\beta Y_\beta + \frac{1}{2} \sum_{\beta=1}^{3} \widehat{c}_\beta \widehat{Y}_\beta,$$

$$\mathcal{L}_{\alpha\beta} = \sum_{\gamma=1}^{3} \epsilon_{\alpha\beta\gamma} a_\gamma, \tag{4.29}$$

and then $a_\beta = c_\beta$, while $\widehat{a}_\beta$ and $\widehat{c}_\beta$ are arbitrary constants.

As for B, we get (here B_S is the symmetric part of B)

$$B = \frac{1}{2} \sum_{\beta=1}^{3} b_\beta Y_\beta + \frac{1}{2} \sum_{\beta=1}^{3} \widehat{b}_\beta \widehat{Y}_\beta + B_S ,$$

$$[B, Y_\alpha] = 0 \Rightarrow b_\alpha = 0, \quad B_S = \lambda I_4 , \tag{4.30}$$

and $\widehat{b}_\beta$ are arbitrary constants.

These results provide a subalgebra of so(8) with basis

$$\begin{pmatrix} Y_\alpha & 0 \\ 0 & Y_\alpha \end{pmatrix}, \quad \begin{pmatrix} \widehat{Y}_\alpha & 0 \\ 0 & 0 \end{pmatrix}, \quad \begin{pmatrix} 0 & 0 \\ 0 & \widehat{Y}_\alpha \end{pmatrix}, \quad \begin{pmatrix} 0 & \widehat{Y}_\alpha \\ \widehat{Y}_\alpha & 0 \end{pmatrix}, \quad \begin{pmatrix} 0 & I_4 \\ -I_4 & 0 \end{pmatrix}. \tag{4.31}$$

In fact, the matrix $\mathcal{L}$ is different from zero only for the generators

$$\begin{pmatrix} Y_\alpha & 0 \\ 0 & Y_\alpha \end{pmatrix}, \tag{4.32}$$

and the corresponding algebra is su(2). The other matrices, which generate a Lie algebra of dimension 10 which commutes with the algebra su(2) generated by the matrices (4.32), leave invariant each of the matrices J_α, $\alpha = 1, 2, 3$.

It is an easy task to construct the adjoint representation and the Killing form and this allows to identify the algebra as a real compact semisimple Lie algebra, sp(2), whose complex extension is isomorphic to the Lie algebra C_2 (or B_2) in the Cartan classification. The details of the computations are given in [70].

If we consider the structure $Y_\alpha^{(i)}$, $i = 2, 3$, the equation to be solved is

$$[X^{(i)}, Y_\alpha^{(i)}] = \sum_{\beta=1}^{3} \mathcal{L}_{\alpha\beta} Y_\beta^{(i)}, \quad \alpha = 1, 2, 3, \ i = 2, 3, \tag{4.33}$$

and then

$$[X^{(i)}, Q_i Y_\alpha^{(1)} Q_i^{-1}] = \sum_{\beta=1}^{3} \mathcal{L}_{\alpha\beta} Q_i Y_\beta^{(1)} Q_i^{-1} \tag{4.34}$$

or

$$[Q_i^{-1} X^{(i)} Q_i, Y_\alpha^{(1)}] = \sum_{\beta=1}^{3} \mathcal{L}_{\alpha\beta} Y_\beta^{(1)}, \tag{4.35}$$

and the algebra formed by the matrices $X^{(i)}$ is now conjugated to the one we get for the first structure $Y_\alpha^{(1)}$. In fact, in the case $Y_\alpha^{(2)}$, the roles of Y_α and $\widehat{Y}_\alpha$ are simply exchanged. The basis for the subalgebra is

$$\begin{pmatrix} \widehat{Y}_\alpha & 0 \\ 0 & \widehat{Y}_\alpha \end{pmatrix}, \ \begin{pmatrix} Y_\alpha & 0 \\ 0 & 0 \end{pmatrix}, \ \begin{pmatrix} 0 & 0 \\ 0 & Y_\alpha \end{pmatrix}, \ \begin{pmatrix} 0 & Y_\alpha \\ Y_\alpha & 0 \end{pmatrix}, \ \begin{pmatrix} 0 & I_4 \\ -I_4 & 0 \end{pmatrix}, \tag{4.36}$$

and the invariance algebra is again sp(2).

Finally, for the third possible structure $Y_\alpha^{(3)}$, we also get sp(2) and a basis is:

$$\begin{pmatrix} Y_\alpha & 0 \\ 0 & \widehat{Y}_\alpha \end{pmatrix}, \ \begin{pmatrix} \widehat{Y}_\alpha & 0 \\ 0 & 0 \end{pmatrix}, \ \begin{pmatrix} 0 & 0 \\ 0 & Y_\alpha \end{pmatrix}, \ \begin{pmatrix} 0 & Z_i \\ -Z_i^T & 0 \end{pmatrix}, \tag{4.37}$$

where Z_i, $i = 1, \ldots, 4$ are the matrices

$$Z_1 = \begin{pmatrix} 1 & 0 & 0 & 0 \\ 0 & 0 & 1 & 0 \\ 0 & -1 & 0 & 0 \\ 0 & 0 & 0 & -1 \end{pmatrix}, \quad Z_2 = \begin{pmatrix} 0 & 1 & 0 & 0 \\ 0 & 0 & 0 & 1 \\ 1 & 0 & 0 & 0 \\ 0 & 0 & 1 & 0 \end{pmatrix},$$

$$Z_3 = \begin{pmatrix} 0 & 0 & 1 & 0 \\ -1 & 0 & 0 & 0 \\ 0 & 0 & 0 & 1 \\ 0 & -1 & 0 & 0 \end{pmatrix}, \quad Z_4 = \begin{pmatrix} 0 & 0 & 0 & -1 \\ 0 & 1 & 0 & 0 \\ 0 & 0 & 1 & 0 \\ -1 & 0 & 0 & 0 \end{pmatrix}, \tag{4.38}$$

which correspond to matrices Q intertwining the two sets Y_α and $\widehat{Y}_\alpha$. The invariance algebra is still the same. This concludes the proof.   △

We have obtained above $L_1 = \mathrm{so}(4) = \mathrm{su}(2) \oplus \mathrm{su}(2)$. The first su(2) corresponds to the strong invariance algebra and the second one to the regular invariance algebra. In the $\mathbf{R}^8$ case, as we have seen, the invariance algebra is $L_2 = \mathrm{su}(2) \oplus \mathrm{sp}(2)$. In fact, the case $\mathbf{R}^4$ has exactly the same structure, since $\mathrm{sp}(1) \approx \mathrm{su}(2)$.

4.5 The General Case: $\mathbf{R}^{4n}$

We are now ready to tackle the general case, i.e. the Euclidean space $\mathbf{R}^{4n}$. We will face two difficulties: the notation, which is necessarily rather cumbersome, and more substantially the identification of the invariance algebra L. In our discussion we will again rely on Lemma 4.1, and hence consider hyperkahler structures in standard form; and will make use of the notation introduced in the discussion of the $\mathbf{R}^8$ case.

Given two square matrices A and B, respectively of dimension m and n, we choose the basis of their tensorial product in such a way that the $m \times m$ block matrix (with $n \times n$ blocks) is

$$A \otimes B = (a_{ij}B), \quad A = (a_{ij}), \tag{4.39}$$

and denote by E_{ij} the usual elementary matrix:

$$(E_{ij})_{kl} := \delta_{ik}\delta_{jl}, \quad E_{ij}E_{kl} = \delta_{jk}E_{il}. \tag{4.40}$$

Then $E_{ij} \otimes B$ is a $nm \times nm$ matrix formed by $n \times n$ blocks, where the ij block is equal to B, and all elements elsewhere are zero.

We recall that the invariance condition for a quaternionic structure in $\mathbf{R}^{4n}$ can be expressed as

$$[X, J_\alpha] = \sum_{\beta=1}^{3} \mathcal{L}_{\alpha\beta}J_\beta, \quad \alpha = 1, 2, 3, \tag{4.41}$$

with $X = -X^T \in \mathcal{M}_{4n}, \mathcal{L} = -\mathcal{L}^T \in \mathcal{M}_3$.

Using Lemma 4.1, any quaternionic structure in $\mathbf{R}^{4n}$ is conjugated to

$$J_\alpha = \sum_{i=1}^{n} E_{ii} \otimes J_\alpha^{(i)}, \quad \alpha = 1, 2, 3, \tag{4.42}$$

where $J_\alpha^{(i)}$ is any of the two nonequivalent quaternionic structures in $\mathbf{R}^4$, i.e. either Y_α or $\widehat{Y}_\alpha$, and we can reduce our problem to the simplest case of block-diagonal quaternionic structures.

Note that we can have in the diagonal both kinds of orientations; this difficulty can be easily surmounted using the fact (already used in the case $\mathbf{R}^8$) that there exist an orthogonal matrix Q, with det $Q = -1$, such that $Y_\alpha = Q^{-1}\widehat{Y}_\alpha Q$, for $\alpha = 1, 2, 3$; see Lemma 4.2 above. We actually extend this to a higher dimensional setting.

Lemma 4.3 *Let J_α be a quaternionic structure in $\mathbf{R}^{4n}$ set in a 4×4 block-diagonal form, $J_\alpha = \sum_{i=1}^{n} E_{ii} \otimes J_\alpha^{(i)}$. Then, the block-diagonal matrix*

$$\mathbf{Q} = \text{diag}(Q_1, Q_2, \ldots, Q_n)\,, \quad Q_i = \begin{cases} I_4 & \text{if } J_\alpha^{(i)} = Y_\alpha, \\ Q & \text{if } J_\alpha^{(i)} = \widehat{Y}_\alpha\,, \end{cases} \qquad (4.43)$$

satisfies

$$\mathbf{Q}^{-1} J_\alpha \mathbf{Q} = \sum_{i=1}^{n} E_{ii} \otimes Y_\alpha. \qquad (4.44)$$

Proof The proof of this is straightforward and hence omitted.   △

We need another preliminary result before going into the explicit computation of the invariance algebra, generalizing Lemmas 4.2 and 4.3; this will allow us to reduce all the different orientation cases to the positively oriented one.

Lemma 4.4 *The invariance algebras of all the quaternionic structures in $\mathbf{R}^{4n}$ are isomorphic.*

Proof We follow, with obvious modifications, the argument used in the case $\mathbf{R}^8$. Thanks to Lemma 4.1, we can just consider structures in standard form. The invariance equation is

$$[X, J_\alpha] = \sum_{\beta=1}^{3} \mathcal{L}_{\alpha\beta} J_\beta, \quad \alpha = 1, 2, 3. \qquad (4.45)$$

If we conjugate the quaternionic structure, $\widetilde{J}_\alpha = U J_\alpha U^{-1}$, we get

$$[X, U^{-1} \widetilde{J}_\alpha U] = \sum_{\beta=1}^{3} \mathcal{L}_{\alpha\beta} U^{-1} \widetilde{J}_\beta U, \quad \alpha = 1, 2, 3; \qquad (4.46)$$

that is,

$$[U X U^{-1}, \widetilde{J}_\alpha] = \sum_{\beta=1}^{3} \mathcal{L}_{\alpha\beta} \widetilde{J}_\beta. \qquad (4.47)$$

Then the algebra generated by the matrices X is isomorphic via a conjugation to the algebra generated by $\widetilde{X} = U X U^{-1}$.

The two operations we make to pass from any quaternionic structure to the positively oriented block-diagonal one, via a preliminary reduction to a block-diagonal one with arbitrary orientation, are conjugations; thus the statement is proved.   △

We are now ready to identify the structure of the invariance algebra L_n. We will split its proof into several lemmas.

Lemma 4.5 *In the case* $(\mathbf{R}^{4n}, I_{4n}, \mathbf{J})$ *the infinitesimal invariance Eq. (4.16) is written as*

$$[X, J_\alpha] = \sum_{\beta=1}^{3} \mathcal{L}_{\alpha\beta} J_\beta, \quad \alpha = 1, 2, 3 \tag{4.48}$$

where $\mathcal{L} = -\mathcal{L}^T \in \mathcal{M}_3$ *is the infinitesimal transformation corresponding to a rotation in* $\mathbf{R}^3$.

Proof In order to preserve the Euclidean metric in $\mathbf{R}^{4n}$, necessarily the infinitesimal transformation $X = -X^T \in \mathcal{M}_{4n}$ belongs to so($4n$). Moreover, Lemma 4.4 guarantees the invariance algebra for different hyperkahler structures on $(\mathcal{V}, g) = (\mathbf{R}^{4n}, I_{4n})$ are isomorphic; thus we only have to study the positively oriented standard one.

That is, we consider the hyperkahler structure given by $\{J_1, J_2, J_3\}$ with

$$J_\alpha = \sum_{i=1}^{n} E_{ii} \otimes Y_\alpha, \quad \alpha = 1, 2, 3. \tag{4.49}$$

It follows that the infinitesimal invariance Eq. (4.16) is given indeed by (4.48), as claimed. $\triangle$

Lemma 4.6 *The subalgebra of* $L_n \subset$ *so($4n$) of the* X *satisfying Eq. (4.48) (and hence leaving invariant the quaternionic structure) has dimension* $n(2n + 1) + 3$, *and a basis of it is provided by*

$$
\begin{pmatrix} Y_\alpha & & & \\ & Y_\alpha & & \\ & & \ddots & \\ & & & Y_\alpha \end{pmatrix}, \quad
\begin{pmatrix} \widehat{Y}_\alpha & & & \\ & 0 & & \\ & & \ddots & \\ & & & 0 \end{pmatrix}, \dots,
\begin{pmatrix} 0 & & & \\ & 0 & & \\ & & \ddots & \\ & & & \widehat{Y}_\alpha \end{pmatrix},
$$

$$
\begin{pmatrix} 0 & \widehat{Y}_\alpha & 0 & \\ \widehat{Y}_\alpha & 0 & 0 & \\ 0 & 0 & 0 & \\ & & & \ddots \\ & & & & 0 \end{pmatrix}, \quad
\begin{pmatrix} 0 & 0 & \widehat{Y}_\alpha & \\ 0 & 0 & 0 & \\ \widehat{Y}_\alpha & 0 & 0 & \\ & & & \ddots \\ & & & & 0 \end{pmatrix},
$$

$$
\dots, \quad
\begin{pmatrix} 0 & & & & \\ & \ddots & & & \\ & & 0 & 0 & 0 \\ & & 0 & 0 & \widehat{Y}_\alpha \\ & & 0 & \widehat{Y}_\alpha & 0 \end{pmatrix},
$$

$$\begin{pmatrix} 0 & I_4 & 0 & & \\ -I_4 & 0 & 0 & & \\ 0 & 0 & 0 & & \\ & & & \ddots & \\ & & & & 0 \end{pmatrix}, \begin{pmatrix} 0 & 0 & I_4 & & \\ 0 & 0 & 0 & & \\ -I_4 & 0 & 0 & & \\ & & & \ddots & \\ & & & & 0 \end{pmatrix},$$

$$\ldots, \begin{pmatrix} 0 & & & & \\ & \ddots & & & \\ & & 0 & 0 & 0 \\ & & 0 & 0 & I_4 \\ & & 0 & -I_4 & 0 \end{pmatrix}.$$

Proof The Eq. (4.48) can be read in the following way. The matrices J_α generate a so(3) algebra. The matrices X obviously form also a subalgebra L of so($4n$), and

$$[[X, \widetilde{X}], J_\alpha] = [[X, J_\alpha], \widetilde{X}] - [[\widetilde{X}, J_\alpha], X] = \sum_\beta \mathcal{L}_{\alpha\beta}[J_\beta, \widetilde{X}] - \sum_\beta \widetilde{\mathcal{L}}_{\alpha\beta}[J_\beta, X]$$

$$= \sum_\gamma [\widetilde{\mathcal{L}}, \mathcal{L}]_{\alpha\gamma} J_\gamma . \tag{4.50}$$

Note that Eq. (4.48) simply states the fact that the subalgebra su(2) generated by the J_α is an ideal of the algebra L.

We can build a basis of so($4n$), which has dimension $2n(4n-1)$, using the matrices $Y_\alpha, \widehat{Y}_\alpha$, their products and the tensor products with the matrices E_{ij}.

More explicitly, such a basis is provided by the matrices

$$A_{ij\alpha} = \frac{1}{2}(E_{ij} + E_{ji}) \otimes Y_\alpha, \quad i, j = 1, \ldots, n, \quad \alpha = 1, 2, 3$$

$$\widehat{A}_{ij\alpha} = \frac{1}{2}(E_{ij} + E_{ji}) \otimes \widehat{Y}_\alpha, \quad i, j = 1, \ldots, n, \quad \alpha = 1, 2, 3$$

$$B_{ij\alpha\beta} = \frac{1}{2}(E_{ij} - E_{ji}) \otimes Y_\alpha \widehat{Y}_\beta, \quad i < j = 1, \ldots, n, \quad \alpha, \beta = 1, 2, 3$$

$$C_{ij} = \frac{1}{2}(E_{ij} - E_{ji}) \otimes I_4, \quad i < j = 1, \ldots, n.$$

It will be notationally convenient to use unconstrained indices i, j with the conventions

$$B_{ij\alpha\beta} = -B_{ji\alpha\beta}, \quad C_{ij} = -C_{ji}, \quad i > j, \quad B_{ii\alpha\beta} = 0, \quad C_{ii} = 0.$$

Some of the commutation relations among these elements will be used in the sequel and can be computed explicitly:

$$[A_{ij\alpha}, A_{kk\gamma}] = \sum_\nu \epsilon_{\alpha\gamma\nu}(\delta_{jk} + \delta_{ik})A_{ij\nu} - \delta_{\alpha\gamma}(\delta_{jk} - \delta_{ik})C_{ij}$$

$$[\widehat{A}_{ij\alpha}, A_{kk\gamma}] = (\delta_{jk} - \delta_{ik})B_{ij\gamma\alpha}$$
$$[B_{ij\alpha\beta}, A_{kk\gamma}] = \sum_{\nu} \epsilon_{\alpha\gamma\nu}(\delta_{jk} + \delta_{ik})B_{ij\nu\beta} - \delta_{\alpha\gamma}(\delta_{jk} - \delta_{ik})\widehat{A}_{ij\beta}$$
$$[C_{ij}, A_{kk\gamma}] = (\delta_{jk} - \delta_{ik})A_{ij\gamma}.$$

Note that the matrices in the positively oriented quaternionic structure are in this notation written as

$$J_\alpha = \sum_i A_{ii\alpha} = \sum_i E_{ii} \otimes Y_\alpha. \tag{4.51}$$

With the convention that indices in the coefficients are also unconstrained, and $b_{ij\alpha\beta} = -b_{ji\alpha\beta}, c_{ij} = -c_{ji}$, the invariance condition for this structure is thus written in terms of the basis (4.51) as

$$\sum_{i,j,k,\alpha} a_{ij\alpha}[A_{ij\alpha}, A_{kk\gamma}] + \sum_{i,j,k,\alpha} \widehat{a}_{ij\alpha}[\widehat{A}_{ij\alpha}, A_{kk\gamma}] + \sum_{i,j,k,\alpha,\beta} b_{ij\alpha\beta}[B_{ij\alpha\beta}, A_{kk\gamma}]$$
$$+ \sum_{i,j,k} c_{ij}[C_{ij}, A_{kk\gamma}] = \sum_{\nu,i} \mathcal{L}_{\gamma\nu} A_{ii\nu}, \quad \gamma = 1, 2, 3. \tag{4.52}$$

Substituting in this the commutators computed above (and understanding all equations are for $\gamma = 1, 2, 3$) we get

$$\sum_{i,j,k,\alpha} a_{ij\alpha,\nu} \epsilon_{\alpha\gamma\nu}(\delta_{jk} + \delta_{ik})A_{ij\nu} - \sum_{i,j,k,\alpha} a_{ij\alpha}\delta_{\alpha\gamma}(\delta_{jk} - \delta_{ik})C_{ij}$$
$$+ \sum_{i,j,k,\alpha} \widehat{a}_{ij\alpha}(\delta_{jk} - \delta_{ik})B_{ij\gamma\alpha} + \sum_{i,j,k,\alpha,\beta,\nu} b_{ij\alpha\beta}\epsilon_{\alpha\gamma\nu}(\delta_{jk} + \delta_{ik})B_{ij\nu\beta}$$
$$- \sum_{i,j,k,\alpha,\beta} b_{ij\alpha\beta}\delta_{\alpha\gamma}(\delta_{jk} - \delta_{ik})\widehat{A}_{ij\beta} + \sum_{i,j,k} c_{ij}(\delta_{jk} - \delta_{ik})A_{ij\gamma} = \sum_{\nu,i} \mathcal{L}_{\gamma\nu} A_{ii\nu};$$

upon standard simplification this reduces to

$$2 \sum_{i,j,\alpha,\nu} \epsilon_{\alpha\gamma\nu} a_{ij\alpha} A_{ij\nu} + 2 \sum_{i,j,\alpha,\beta,\nu} \epsilon_{\alpha\gamma\nu} b_{ij\alpha\beta} B_{ij\nu\beta} = \sum_{\nu,i} \mathcal{L}_{\gamma\nu} A_{ii\nu}. \tag{4.53}$$

This should be seen as a matrix equation, i.e. a set of scalar equations, for the coefficients a, b (note the c_{ij} cancelled out) and for the matrix elements $\mathcal{L}_{ij}$. As for the a_{ijk} and the $\mathcal{L}_{ij}$, the solution is

$$a_{ij\alpha} = 0, \quad i \neq j \tag{4.54}$$
$$\mathcal{L}_{12} = 2a_{ii3}, \quad \mathcal{L}_{13} = -2a_{ii2}, \quad \mathcal{L}_{23} = 2a_{ii1}. \tag{4.55}$$

On the other hand the equations for $b_{ij\alpha\beta}$ have the unique solution

$$b_{ij\alpha\beta} = 0, \quad i, j = 1, \ldots, n, \ \alpha, \beta = 1, 2, 3. \tag{4.56}$$

The invariance algebra is then formed by the elements of the form

$$\begin{aligned}
X &= \sum_{i,j,\alpha} a_{ij\alpha} A_{ij\alpha} + \sum_{i,j,\alpha} \widehat{a}_{ij\alpha} \widehat{A}_{ij\alpha} + \sum_{i,j} c_{ij} C_{ij} \\
&= \frac{1}{2}\mathcal{L}_{23} J_1 + \frac{1}{2}\mathcal{L}_{31} J_2 + \frac{1}{2}\mathcal{L}_{12} J_3 \\
&\quad + \frac{1}{2}\sum_{i,j,\alpha} \widehat{a}_{ij\alpha}(E_{ij} + E_{ji}) \otimes \widehat{Y}_\alpha + \frac{1}{2}\sum_{i,j} c_{ij}(E_{ij} - E_{ji}) \otimes I_4. \tag{4.57}
\end{aligned}$$

One can check explicitly that these elements satisfy the invariance equation.

It seems at first sight that this would leave open the possibility that other elements are also in the invariance algebra. But actually the algebra spanned by these elements is a maximal subalgebra of so($4n$); given that obviously not all elements of so($4n$) preserve the quaternionic structure, one is guaranteed to have indeed identified the full invariance algebra. $\triangle$

Theorem 4.3 *The invariance algebra L_n is the direct sum of two mutually commuting subalgebras,*

$$L = \mathrm{su}(2) \oplus \mathcal{G}, \tag{4.58}$$

one of them being the su(2) *algebra generated by the J_α, and the other being a Lie algebra of dimension $n(2n + 1)$.*

Proof Obviously there is a subalgebra generated by the three first diagonal elements in (4.57), i.e. by $X_\alpha = \mathrm{diag}(Y_\alpha, \ldots, Y_\alpha)$; this is precisely the su(2) subalgebra.[1] It is clear from (4.57) that all other elements also form a subalgebra, and that the two subalgebras commute due to $[Y_\alpha, \widehat{Y}_\beta] = 0$. The statement on the dimension of $\mathcal{G}$ follows by direct inspection. $\triangle$

It also follows easily from $[Y_\alpha, \widehat{Y}_\beta] = 0$ that $\mathcal{G}$ is actually the strong invariance algebra for **J**.

We are left with the task of identifying the Lie algebra $\mathcal{G}$; this is not immediate and will require some Lie algebra theory. We actually know that $\mathcal{G}$ is equal to sp(n) when $n = 1, 2$, see Theorems 4.1 and 4.2. It turns out that this is always the case.

Theorem 4.4 *The strong invariance algebra for the standard positively oriented hyperkahler structure on Euclidean $\mathbf{R}^{4n}$ is $\mathcal{G} = \mathrm{sp}(n)$.*

Proof Let us consider the complex extension of $\mathcal{G}$. We can construct a Chevalley basis following the same procedure as in the case $\mathbf{R}^8$. We first define

$$H = i\widehat{Y}_3, \quad E_+ = \frac{1}{2}(\widehat{Y}_1 + i\widehat{Y}_2), \quad E_- = \frac{1}{2}(-\widehat{Y}_1 + i\widehat{Y}_2) \tag{4.59}$$

---

[1] This fact corresponds to the one, already remarked, that (4.48) means that the subalgebra generated by the J_α is an ideal in L.

with commutation relations

$$[H, E_+] = 2E_+, \quad [H, E_-] = -2E_-, \quad [E_+, E_-] = H. \tag{4.60}$$

Using these, we define new matrices, which are linear combinations of the elements defined above:

$$
\begin{aligned}
\mathcal{H}_i &= (E_{ii} - E_{i+1,i+1}) \otimes H, \quad i = 1, \ldots, n-1 \\
\mathcal{H}_n &= E_{nn} \otimes H \\
\mathcal{E}^\ell_{\pm,j} &= E_{jj} \otimes E_\pm, \quad j = 1, \ldots, n \\
\mathcal{E}^{s,1}_{\pm,jk} &= \pm \frac{1}{2}(E_{jk} - E_{kj}) \otimes I_4 + \frac{1}{2}(E_{jk} + E_{kj}) \otimes H, \quad 1 \le j < k \le n \\
\mathcal{E}^{s,2}_{\pm,jk} &= (E_{jk} + E_{kj}) \otimes E_\pm, \quad 1 \le j < k \le n.
\end{aligned}
\tag{4.61}
$$

It turns out that, as can be checked by an explicit computation, the matrices

$$\mathcal{H}_i, \quad \mathcal{E}^\ell_{\pm,j}, \quad \mathcal{E}^{s,1}_{\pm,jk}, \quad \mathcal{E}^{s,2}_{\pm,jk} \tag{4.62}$$

form a basis of the Lie algebra C_n (in the Cartan notation) in a $4n$-dimensional representation. The matrices $\mathcal{H}_i$ ($i = 1, \ldots, n$) are a basis of a Cartan subalgebra; the matrices $\mathcal{E}^\ell_{\pm,j}$ are the root vectors corresponding to the long roots, and the $\mathcal{E}^{s,r}_{\pm,jk}$ to the short ones. The details of the computation of the commutation relations, and in particular the determination of the root system, which show that this is the Lie algebra C_n, are given in [70]. △

We are now ready to state and proof our main result.

Theorem 4.5 *The invariance algebra for* $(\mathbf{R}^{4n}, I_{4n}, \mathbf{J})$ *is*

$$L_n = su(2) \oplus sp(n).$$

Proof We come back considering $L = su(2) \oplus \mathcal{G} \subset so(4n)$. The complex extension of the orthogonal algebra $so(4n)$, is D_{2n} in the Cartan notation [93]. The maximal subgroups of the classical groups were classified by Dynkin in [53], and the result we need is that $A_1 \oplus C_n$ is a maximal subalgebra of the Lie algebra D_{2n}, which is in agreement with our results in Theorems 4.3 and 4.4.

In fact the fundamental representation $(10 \cdots 0)$, in the highest weight notation, of D_{2n} is irreducible when restricted to the subalgebra $A_1 \oplus C_n$, as

$$D_{2n} \to A_1 \oplus C_n, \quad (1\overbrace{0\cdots0}^{2n-1}) \to (1)(1\overbrace{0\cdots0}^{n-1}), \tag{4.63}$$

and decomposes, when restricted to A_1, into $2n$ copies of the spin 1/2 representation (i.e. (1) in the highest weight notation),

$$D_{2n} \rightarrow A_1, \quad (1\overset{2n-1}{\overbrace{0\cdots0}}) \rightarrow 2n(1); \tag{4.64}$$

and, when restricted to C_n, into the sum of two copies of the fundamental representation of C_n:

$$D_{2n} \rightarrow C_n, \quad (1\overset{2n-1}{\overbrace{0\cdots0}}) \rightarrow 2(1\overset{n-1}{\overbrace{0\cdots0}}). \tag{4.65}$$

Since $\mathcal{G}$ is a semisimple compact Lie algebra, we finally have the complete structure of the algebra L, i.e. $L = \mathrm{su}(2) \oplus \mathrm{sp}(n)$, as claimed.   △

Our discussion shows, as mentioned in passing, that the invariance algebra $L_n = \mathrm{su}(2) \oplus \mathcal{G} \subset \mathrm{so}(4n)$ is actually a maximal subalgebra of $\mathrm{so}(4n)$. The elements in $\mathrm{su}(2)$ correspond to regular invariance transformations, and those in $\mathcal{G} = \mathrm{sp}(n)$ to the strong invariance ones.

We should note that although the invariance algebras are isomorphic for different quaternionic structures on $\mathbf{R}^{4n}$, their realizations are not the same and depend on the different quaternionic structures (in particular, on the different orientations they may have). We have seen in the case $\mathbf{R}^8$ how they can be constructed and the differences among them. The construction follows essentially the same lines in the general case.

Remark 4.2 The appearance of the (compact) symplectic algebra $\mathrm{sp}(n)$ is not a surprising result in this context. In fact, it can be identified with the Lie algebra $\mathrm{sl}(n, \mathbf{H})$ of quaternionic $n \times n$ matrices with purely imaginary trace [17]. The symplectic group $\mathrm{Sp}(2n)$ can be realized in terms of quaternions as a subgroup of the general linear group $\mathrm{GL}(n, \mathbf{H})$.   ⊙

Chapter 5
Integrable Hyperhamiltonian Systems

Among all possible hyperhamiltonian dynamics, a special role is played by *integrable* ones [59, 61, 69]; these also turn out to be associated to relevant physical applications.

Note that—albeit it is clear that an integrable system should be a system which can be explicitly integrated—we will also need a formal definition of integrability; as usual we will provide this along the lines of the standard Hamiltonian one, but the different structures at hand will make that some relevant differences are present. In particular, in the same way as in Hamiltonian dynamics any integral of motion "counts for two" in the reduction process, here any integral of motion will "count for four" (the meaning of this will be made precise in Sect. 5.2).

In this chapter we will always deal with Euclidean $\mathbf{R}^{4n}$ spaces.

5.1 Quaternionic Oscillators

The prototypical Hamiltonian integrable system is the *harmonic oscillator*, or a system thereof (in fact, Arnold-Liouville integrable systems can also be defined as systems which can be mapped into an oscillator system [73]). In our case, the analogue of these will be provided by *quaternionic oscillators*.

5.1.1 Integrable Hamiltonian Systems

We start by very briefly recalling the main features of standard oscillator systems and (Arnold-Liouville) Hamiltonian integrable systems [10, 73].

© Springer International Publishing AG 2017
G. Gaeta and M.A. Rodríguez, *Lectures on Hyperhamiltonian
Dynamics and Physical Applications*, Mathematical Physics Studies,
DOI 10.1007/978-3-319-54358-1_5

A Hamiltonian system in n degrees of freedom—which, we recall, is a system of $2n$ first order ODEs—is *integrable* if it can be mapped (via a diffeomorphism) to

$$\dot{q}_k = \nu_k \, p_k \, , \quad \dot{p}_k = -\nu_k \, q_k \quad (k = 1, \ldots, n) \tag{5.1}$$

that is, to an n-dimensional harmonic oscillator.

Here the ν_k only depend (possibly) on the quantities $I_k = (p_k^2 + q_k^2)$. Note that there is no sum on repeated indices, and the same will apply in all of this subsection.

Passing to action-angle coordinates (I, φ),

$$p_k = \sqrt{I_k} \, \cos \varphi_k \, , \quad q_k = \sqrt{I_k} \, \sin \varphi_k \tag{5.2}$$

(this transformation is singular in the origin[1]), the evolution reads

$$\dot{I}_k = 0 \, , \quad \dot{\varphi}_k = \nu_k \, . \tag{5.3}$$

Equivalently, one can use complex coordinates

$$z_k = p_k + i \, q_k \equiv \sqrt{I_k} \, \exp[i \, \varphi_k] \, ; \tag{5.4}$$

in these the evolution reads

$$\dot{z}_k = i \, \nu_k \, z_k \, . \tag{5.5}$$

We stress again that the frequencies ν_k are in general a function of the I_k (or equivalently of the $|z_k|^2$) variables.

Remark 5.1 It should be stressed that the system is invariant under the abelian group

$$SO(2) \times \cdots \times SO(2) \equiv U(1) \times \ldots \times U(1) = \mathbf{T}^n \, ; \tag{5.6}$$

moreover, time evolution is given by a (real or complex) rotation, with speed ν_k, in each ($\mathbf{R}^2$ or $\mathbf{C}^1$) subspace. $\odot$

Remark 5.2 Conversely, a system enjoying a $\mathbf{T}^n$ symmetry is integrable (in Arnold-Liouville sense) [10], and it is obvious that if the phase space can be fibred in terms of two-dimensional manifolds M^2 as $M^2 \times \ldots \times M^2$, or in terms of complex lines $\mathbf{C}^1$ as $\mathbf{C}^1 \times \ldots \times \mathbf{C}^1$, with time evolution described by rotations in each factor, then the system is integrable. $\odot$

---

[1]The inverse transformation is $I_k = p_k^2 + q_k^2$, $\varphi_k = \arctan(q_k/p_k)$, which is thus singular on the whole line $p_k = 0$.

5.1.2 Quaternionic Oscillators

The idea to deal with quaternionic oscillators (and hence integrable quaternionic dynamics) will be to replace complex rotations (in $\mathbf{C}^1 \simeq \mathbf{R}^2$ spaces) with *quaternionic rotations* (in $\mathbf{H}^1 \simeq \mathbf{R}^4$ spaces).

Definition 5.1 A *simple quaternionic oscillator* in the Euclidean $\mathbf{R}^4$ space is a four-dimensional system of first order ODEs of the form

$$\dot{x}^i = \sum_{\alpha=1}^{3} c_\alpha(|\mathbf{x}|^2) \, (L_\alpha)^i_{\ j} \, x^j \,, \tag{5.7}$$

where of course $|\mathbf{x}|$ is the norm of the vector $\mathbf{x} = (x^1, x^2, x^3, x^4)$, and the L_α are real four-dimensional matrices satisfying the quaternionic relations (1.15).

The *frequency* of the oscillator is

$$\nu = \nu(|\mathbf{x}^2|) = \sqrt{c_1^2(|\mathbf{x}|^2) + c_2^2(|\mathbf{x}|^2) + c_3^2(|\mathbf{x}|^2)} \,. \tag{5.8}$$

If the c_α are constant,[2] we can always reduce to a Hamiltonian system; note that we could have constant ν even with non-constant c_α, thus a constant frequency does *not* imply the system can be reduced to a Hamiltonian one.

This definition is readily extended to the higher dimensional case, i.e. to systems in (Euclidean) $\mathbf{R}^{4n}$. In view of this, it is convenient to introduce some compact notation. We consider coordinates $(x^1, \ldots, x^{4n})$ in $M = \mathbf{R}^{4n}$; together with x, we consider a set of n four-dimensional vectors $\xi_{(1)}, \ldots \xi_{(n)}$ with components

$$\xi^i_{(k)} = x^{4(k-1)+i} \,. \tag{5.9}$$

In other words, we have decomposed the $4n$-dimensional vector into subvectors $\xi_{(k)}$ of dimension four,

$$\mathbf{x} = \left((\xi_{(1)}), \ldots, (\xi_{(n)}) \right) = \left(\xi^1_{(1)}, \xi^2_{(1)}, \ldots, \xi^3_{(n)}, \xi^4_{(n)} \right) \,.$$

Definition 5.2 A general *quaternionic oscillator* in $\mathbf{R}^{4n}$ is a $4n$-dimensional system of first order ODEs of the form

$$\dot{\xi}^i_{(k)} = \sum_{\alpha=1}^{3} c_\alpha^{(k)}(|\xi_{(1)}|^2, \ldots, |\xi_{(n)}|^2) \, \left(Y_\alpha^{(k)} \right)^i_{\ j} \, \xi^j_{(k)} \,, \tag{5.10}$$

where the $Y_\alpha^{(k)}$ are real four-dimensional matrices satisfying the quaternionic relations (1.15).

---

[2]If the c_α are constant functions, we say we have a *linear* quaternionic oscillator.

The *frequencies* ν_k $(k = 1, \ldots, n)$ of the oscillator are

$$\nu_{(k)} = \nu_k(|\xi_{(1)}|^2, \ldots, |\xi_{(n)}|^2) = \sqrt{\sum_{\alpha=1}^{3}[c_\alpha^{(k)}(|\xi_{(1)}|^2, \ldots, |\xi_{(n)}|^2)]^2} . \tag{5.11}$$

Remark 5.3 Quaternionic oscillators can of course also be defined without reference to the $\xi_{(i)}$ vectors; we have then

$$\dot{x}^i = \sum_\alpha c_\alpha[(\mathbf{x} \cdot B_1\mathbf{x}), \ldots, (\mathbf{x} \cdot B_n\mathbf{x})] \, (L_\alpha)^i{}_j \, x^j ; \tag{5.12}$$

here B_k is a sparse matrix having as only nonzero elements those on the diagonal at positions from $(4(k-1)+1)$ to $4k$, $(\mathbf{x} \cdot B_k\mathbf{x})$ denotes the scalar product between the vectors $\mathbf{x}$ and $B_k\mathbf{x}$, the c_α are functions of the quantities $(\mathbf{x} \cdot B_k\mathbf{x})$ $(k = 1, \ldots, n)$; and the L_α are block-diagonal matrices with four-dimensional blocks, satisfying the quaternionic relations. These are now written in the form

$$L_\alpha L_\beta = \sum_\gamma \epsilon_{\alpha\beta\gamma} L_\gamma - \delta_{\alpha\beta} I_{4n} , \tag{5.13}$$

with I_{4n} the $4n$-dimensional identity matrix.   ⊙

Remark 5.4 Writing, with an obvious notation, the $4n$-dimensional (and block-reducible) matrices L_α as

$$L_\alpha = L_\alpha^{(1)} \oplus \ldots \oplus L_\alpha^{(n)} , \tag{5.14}$$

it is immediate to see that the sub-matrices corresponding to each four-dimensional block also satisfy the quaternionic relations (5.13). Hence we conclude that acting on $\mathbf{R}^{4n}$ via a linear transformation in

$$SO(4) \times \ldots \times SO(4) \subset SO(4n) ,$$

in each block the matrices $L_\alpha^{(k)}$ can be reduced to standard ones, with either positive or negative orientation depending on the matrices L_α, say with $m = 0, \ldots, n$ positively oriented and $n - m$ negatively oriented blocks. Note also that by a (block) permutation of variables (which does not alter the orientation in $\mathbf{R}^{4n}$), we can always reduce to the case where the first m blocks have positive orientation, and the remaining $(n - m)$ have negative one.   ⊙

5.2 Action-Spin Coordinates

In this section we discuss the integration of quaternionic oscillators. In the same way as integration of harmonic oscillators in standard Hamiltonian dynamics is intimately related to action-angle coordinates, the integration of quaternionic oscillators is related to *action-spin* coordinates.

5.2.1 The Simple Quaternionic Oscillator Dynamics

We can now proceed to integrate (5.10); we will actually start from (5.7), for ease of discussion. We rewrite this in the form

$$\dot{x}^i = L^i{}_j x^j , \tag{5.15}$$

where of course, compare (5.7),

$$L = \sum_\alpha c_\alpha(|\mathbf{x}|^2) L_\alpha .$$

Recall now that L_α, and hence L, are skew-symmetric. Then $\rho = |\mathbf{x}|^2$ evolves according to

$$\frac{d\rho}{dt} = \frac{d|\mathbf{x}|^2}{dt} = 2\,x^i \frac{dx^i}{dt} = 2\,x^i L^i{}_j x^j = 0 ; \tag{5.16}$$

that is, $\rho = |\mathbf{x}|^2$ is a constant of motion. Therefore,

$$c_\alpha(|\mathbf{x}|^2) \equiv c_\alpha(\rho)$$

are also constant along the motion, and are identified by their initial values.

It follows from this that L is also constant on the dynamics. Thus the solution to (5.15) is simply written as

$$x(t) = \exp[Lt]\,x(0) . \tag{5.17}$$

Here of course the exponential has to be meant in the sense of the series expansion, i.e.

$$e^{Lt} = I + t L + (t^2/2) L^2 + \dots = \sum_{m=0}^{\infty} \frac{t^m}{m!} L^m . \tag{5.18}$$

Our problem is thus to evaluate this series.

As for L^2, we have (with I_4 the four-dimensional identity matrix)

$$L^2 = \left(\sum_\alpha c_\alpha L_\alpha \right) \left(\sum_\beta c_\beta L_\beta \right) = \sum_{\alpha,\beta} c_\alpha \, c_\beta \left[\epsilon_{\alpha\beta\gamma} \, L_\gamma \, - \, \delta_{\alpha\beta} I_4 \right]$$

$$= - \left(\sum_\alpha c_\alpha^2 \right) I_4 := - \nu^2 \, I_4 \ .$$

Now we note that ν^2 depends on the x^i only through ρ. Again by (5.16), it follows that ν^2, and hence $\nu = \sqrt{\nu^2}$, are constants of motion.

As ν^2 is a constant, and taking also into account $L^2 = -\nu^2 I_4$, we have, going back to (5.17) and recalling (5.18), that

$$\exp[Lt] = \sin(\nu t) \, L + \cos(\nu t) \, I_4 \ . \tag{5.19}$$

With $\mathbf{x}_0 = \mathbf{x}(0)$ the initial condition, the solution (5.17) reads

$$x(t) = [\cos(\nu t) \, I_4 + \sin(\nu t) \, L] \, \mathbf{x}_0 \ . \tag{5.20}$$

Remark 5.5 The solutions (5.20) live on the sphere S^3 of radius $\rho_0 = |\mathbf{x}_0|$ (as obvious from $d\rho/dt = 0$), moving on great circles S^1 identified by $\mathbf{x}_0$ and $\mathbf{x}_1 = L\mathbf{x}_0$. They realize the Hopf fibration of S^3 [94]. We stress that—albeit we have not written this explicitly for ease of notation—in (5.19) and (5.20) ν is a function of ρ; thus (contrary to what happens for harmonic oscillators) it takes in general different values on different spheres.[3]   ⊙

Remark 5.6 A special type of hyperhamiltonian system is provided by standard Hamiltonian ones. The system (5.7) is Hamiltonian if only one of the $\mathcal{H}^\alpha$, say $\mathcal{H}^1$, is nonzero. In this case we actually have *two* global constants of motion: e.g., if the only nonzero Hamiltonian is $\mathcal{H}_1$ and $L_\alpha = Y_\alpha$, these are $I_1 = (x^1)^2 + (x^2)^2$ and $I_2 = (x^3)^2 + (x^4)^2$. Recall that for genuinely hyperhamiltonian systems we have only $\rho = |\mathbf{x}|^2$. It should be stressed that, as we have a closed curve in a four-dimensional space, we always have three constants of motion; the other two will depend on the radius ρ of the sphere (i.e. on the initial datum), or more precisely on the values taken by $c_\alpha(\rho)$ on these spheres.   ⊙

---

[3]The closest analogue is thus an oscillator with amplitude-depending frequency $\omega = \omega(I)$. For ν a constant we have a straight equivalent of standard (isochronous) oscillators, $\omega = \omega_0$; we will see in a moment this case is actually "too closely" equivalent to a standard oscillator.

5.2.2 Action-Spin Coordinates

The equivalent of action-angle coordinates are now *action-spin* coordinates (I, s^{α}), where now $I = |\mathbf{x}|^2 \in \mathbf{R}$, and the s^{α} are coordinates in $SU(2) \simeq S^3$ (as the sphere S^3 is parallelizable [94], these are global coordinates). This also shows that quaternionic oscillators describe an evolution on the $SU(2)$ group, governed by an element of the $su(2)$ Lie algebra which depends only on $\rho = |\mathbf{x}|^2 = I$ and is hence constant on the level manifolds for I (i.e. on spheres S^3 of given radius).[4]

5.2.3 Generic Quaternionic Oscillators

The case of generic quaternionic oscillators can be discussed along the same lines. We now define

$$\rho_k := |\xi_{(k)}|^2 \; ;$$

each of these $(k = 1, \ldots, n)$ is a constant of motion (as follows again from the skew-symmetry of the relevant matrices), and hence the c_{α} in (5.10) are constant on the dynamics. Moreover, the matrices L_{α} are block-reducible, $L_{\alpha} = L_{\alpha}^{(1)} \oplus \ldots \oplus L_{\alpha}^{(k)}$. Thus

$$\exp[t L_{\alpha}] = \exp[t L_{\alpha}^{(n)}] \oplus \ldots \oplus \exp[t L_{\alpha}^{(n)}] \, ,$$

and each $\exp[t L_{\alpha}^{(k)}]$ is computed in the same way as shown above for simple quaternionic oscillators.

We thus reach the same conclusion for the solution issued from an initial datum $x(0) = x_0$ (now with $x_0 \in \mathbf{R}^{4n}$), i.e. it is given by

$$x(t) = [\cos(\nu t) \, I_{4n} + \sin(\nu t) \, L] \, x_0 \, . \tag{5.21}$$

Now we have action-spin coordinates $(\mathbf{I}, s^{\alpha})$ with $\mathbf{I} = (I_{(1)}, \ldots, I_{(n)})$, and similarly $s^{\alpha} = (s_{(1)}^{\alpha}, \ldots, s_{(n)}^{\alpha})$. Here the actions $I_{(k)} = |\xi_{(k)}|^2$ are constants of motion, while $s_{(k)}$ are coordinates on $SU(2) \simeq S^3$. It should be stressed that we may in principles have a different set of matrices—i.e. a different $su(2)$ representation—in each block.[5]

Remark 5.7 It is well known that in a generic dynamical system of dimension m (i.e. a system of m first order ODEs), knowledge of a constant of motion allows to reduce

---

[4]The concrete integration of a (linear) motion on $SU(2)$, or more generally on Lie groups, is performed by the well known Wei-Norman method [33, 145, 146].

[5]Actually, as discussed in Remark 5.6 in Chap. 1 (see Sect. 1.4.3), these can always be reconducted to the standard ones, so only the orientation is actually arbitrary. By a permutation of the blocks, we can always have first blocks with $su(2)$ representations having the positive orientation, and then those with the negative one.

the dimension of the system by one, i.e. reducing the system to one of dimension $m - 1$. On the other hand, if the system under study is Hamiltonian (or more generally has a variational nature - so this also holds for Lagrangian systems) with n degrees of freedom (i.e. a system of $m = 2n$ first order equations with a Hamiltonian structure) knowledge of a constant of motion I_1 allows to reduce the system with one of $n - 1$ degrees of freedom (i.e. a system of $m' = 2(n - 1)$ first order ODEs). In this sense, for Hamiltonian systems each constant of motion "counts for two". This is due to the fact that passing to coordinates such that I_1 is one of the variables, the conjugate one (the angle ϑ_1 related to the action I_1) is also easily determined.

In the case of hyperhamiltonian systems, a similar feature is present, but now there are *three* variables conjugate to the action I_1, and we have just shown how their evolution can be determined if I_1 is known to be constant. Thus knowledge of a constant of motion in a hyperhamiltonian system with n degrees of freedom (i.e. a system of $m = 4n$ first order ODEs with a hyperkahler structure) allows to reduce to a hyperhamiltonian system with $n - 1$ degrees of freedom, i.e. a system of $m' = 4(n - 1)$ first order ODEs. In this sense for hyperhamiltonian systems each constant of motion "counts for four", as stated in the first lines of this Chapter.   ⊙

5.3   Are Hyperhamiltonian Integrable Systems More General Than Hamiltonian Ones?

We have seen in Sect. 2.2 that hyperhamiltonian dynamics is an effective extension of standard Hamiltonian one, in that it contains the latter as a proper subset.

It would be however possible, in principles, that the inclusion is not proper in the integrable case; that is, one may wonder if really there are integrable hyperhamiltonian systems which are not Hamiltonian.[6]

The purpose of this section is to show that this is indeed the case; our discussion will follow the one given in [59], and proceed by an explicit example.

We consider $M = \mathbf{R}^4$ with Euclidean metric and standard hyperkahler structure; we will use standard cartesian coordinates x^i ($i = 1, \ldots, 4$) on M and write $\rho = |\mathbf{x}|^2$. As for the Hamiltonians, we will choose

$$\mathcal{H}^1 = \rho \; ; \quad \mathcal{H}^2 = \rho^2/2 \; ; \quad \mathcal{H}^3 = 0 \; . \tag{5.22}$$

We will write, for later reference, $h_\alpha = d\mathcal{H}^\alpha/d\rho$; thus with our choice we get

$$h_1 = 1 \; , \quad h_2 = \rho \; , \quad h_3 = 0 \; . \tag{5.23}$$

---

[6]Note, in this respect, that if the c_α are constant, (n-dimensional) quaternionic oscillators can be reduced to ($2n$-dimensional) standard ones; this also means that *on a given sphere* (but, in general, not in any neighborhood of this) quaternionic oscillators can be recast in standard Hamiltonian form. Thus isochronous quaternionic oscillators are "not interesting" in that they are always Hamiltonian.

With our choices, the hyperhamiltonian flow is given by

$$\dot{x} = \sum_\alpha Y_\alpha \nabla \mathcal{H}^\alpha = \sum_\alpha h_\alpha Y_\alpha x ; \tag{5.24}$$

in fully explicit terms, we get

$$\dot{x}^i = (Y_1 + \rho Y_2)^i{}_j x^j . \tag{5.25}$$

We should now ascertain if this, or equivalently the associated VF

$$X = [(Y_1 + \rho Y_2)^i{}_j x^j] \partial_i , \tag{5.26}$$

can be written in Hamiltonian terms for *some* symplectic structure ω and Hamiltonian H.

The symplectic structure, whatever it is, will be written in coordinates as

$$\omega = \frac{1}{2} A_{ij} \, dx^i \wedge dx^j \tag{5.27}$$

for some skew-symmetric matrix A (this is also subject to the constraints corresponding to $d\omega = 0$; we will see about this in a moment). We then have

$$X \lrcorner \omega = \sum_\alpha [Y_\alpha (\nabla \mathcal{H}^\alpha)]^i A_{ij} \, dx^j = [(Y_1 + \rho Y_2)^i{}_k x^k] A_{ij} \, dx^j := W_j \, dx^j ; \tag{5.28}$$

here we have of course defined

$$W_i = \sum_\alpha [Y_\alpha (\nabla \mathcal{H}^\alpha)]^k A_{ki} = [(Y_1 + \rho Y_2)^k{}_m x^m] A_{ki} . \tag{5.29}$$

If X is Hamiltonian w.r.t. the symplectic structure ω, there exists a smooth scalar function $H : \mathbf{R}^4 \to \mathbf{R}$ such that

$$X \lrcorner \omega = dH . \tag{5.30}$$

As we are in $\mathbf{R}^4$, this is equivalent to

$$d(X \lrcorner \omega) = 0 ; \tag{5.31}$$

this is in turn equivalent, using our coordinate expressions, to

$$P_{ij} := \partial_i W_j - \partial_j W_i = 0 . \tag{5.32}$$

This should be seen as an equation for the coefficients A_{ij} of the symplectic form ω, while the Y_α and the Hamiltonians $\mathcal{H}^\alpha$ are given.

The only apriori information we have on A is that it is a skew-symmetric matrix, and that it is constant on spheres. Thus we can write

$$A = \begin{pmatrix} 0 & a_1(\rho) & a_2(\rho) & a_3(\rho) \\ -a_1(\rho) & 0 & a_4(\rho) & a_5(\rho) \\ -a_2(\rho) & -a_4(\rho) & 0 & a_6(\rho) \\ -a_3(\rho) & -a_5(\rho) & -a_6(\rho) & 0 \end{pmatrix}, \tag{5.33}$$

with $a_k(\rho)$ unknown smooth functions.

We must now proceed to determine the functions $a_k(\rho)$ so that (5.32) are satisfied. This is a straightforward matter and we omit details; requiring $P_{12} = 0$ yields $a_2(\rho) = a_5(\rho) = 0$, $a_3(\rho) = \rho a_1(\rho)$, $a_4(\rho) = \rho a_1(\rho)$. Next we consider P_{14}, and setting this to zero just requires $a_6(\rho) = 0$. At this point (5.32) are satisfied, and we are left with

$$A = a_1(\rho) \begin{pmatrix} 0 & 1 & 0 & \rho \\ -1 & 0 & \rho & 0 \\ 0 & -\rho & 0 & 1 \\ -\rho & 0 & -1 & 0 \end{pmatrix}. \tag{5.34}$$

To this matrix A corresponds the form

$$\omega = a_1(\rho) \left[\omega_1 + \rho \omega_2 \right], \tag{5.35}$$

which is non-degenerate provided $a_1(\rho)$ is nowhere zero.

Note, however, that we cannot yet claim this is a symplectic form: in fact we have not taken care of the requirement that ω is closed (this would imply $\epsilon_{ijk}\partial_i A_{jk} = 0$, and we have not enforced this condition on A). At this point we can work directly on ω (rather than on A), and by simple algebra (using the fact a_1 only depends on x through ρ) we get

$$\begin{aligned} d\omega = & \left[x_3 a_1' + x_1(a_1 + \rho a_1') \right] dx^1 \wedge dx^2 \wedge dx^3 \\ & + \left[x_4 a_1' - x_2(a_1 + \rho a_1') \right] dx^1 \wedge dx^2 \wedge dx^4 \\ & + \left[x_1 a_1' - x_3(a_1 + \rho a_1') \right] dx^1 \wedge dx^3 \wedge dx^4 \\ & + \left[x_2 a_1' + x_4(a_1 + \rho a_1') \right] dx^2 \wedge dx^3 \wedge dx^4. \end{aligned}$$

Taking again into account that $a_1 = a_1(\rho)$, it is clear that these imply $a_1 = 0$, and hence $\omega = 0$.

We conclude that in this case the hyperhamiltonian vector field X cannot be described in Hamiltonian terms, for *any* symplectic structure. This shows, by a concrete and fully explicit example, that there are hyperhamiltonian integrable systems which are *not* Hamiltonian.

Remark 5.8 It may be worth discussing our procedure in a more general setting, with a view at identifying hyperhamiltonian integrable systems which can (or cannot) be set in standard Hamiltonian form. Proceeding as above but with unspecified Hamiltonians $\mathcal{H}^\alpha$, we get immediately

$$
\begin{aligned}
P_{ij} &= \left[\partial_i A_{kj} - \partial_j A_{ki}\right] M_\alpha^{k\ell} (\partial_\ell \mathcal{H}^\alpha) + M_\alpha^{k\ell} \left[(\partial_{i\ell}^2 \mathcal{H}^\alpha) A_{kj} - (\partial_{j\ell}^2 \mathcal{H}^\alpha) A_{ki}\right] \\
&= (\partial_k A_{ij}) M_\alpha^{k\ell} (\partial_\ell \mathcal{H}^\alpha) + \left[A_{ik} M_\alpha^{k\ell} (\partial_{\ell j}^2 \mathcal{H}^\alpha) - (\partial_{i\ell}^2 \mathcal{H}^\alpha) M_\alpha^{\ell k} A_{kj}\right] ;
\end{aligned}
$$

here in the last step we have used $A^T = -A$, $M_\alpha^T = -M_\alpha$, while in the previous one we have used the assumption that $d\omega = 0$, which –as already mentioned –implies $\epsilon_{ijk} \partial_i A_{jk} = 0$ (and again skew-symmetry of A). The vanishing of these for all $i \neq j$ are the condition for setting the system in Hamiltonian form.

If we assume that $\mathcal{H}^\alpha(x) = f_\alpha(\rho)$ (we write the index α as a lower one for ease of notation, and take $\varrho = |\mathbf{x}|^2/2 = (1/2)\rho$ for the same reason), then these read

$$
\begin{aligned}
P_{ij} &= f_\alpha' \left[(\partial_k A_{ij}) (M_\alpha)^{k\ell} x_\ell + A_{ik} (Y_\alpha)^k{}_j - (Y_\alpha)_i{}^k A_{kj}\right] \\
&+ f_\alpha'' x_\ell M_\alpha^{\ell k} \left[A_{ki} x_j - A_{kj} x_i\right] ,
\end{aligned}
$$

and the vanishing of P_{ij} is the necessary and sufficient condition for setting X in Hamiltonian form. $\odot$

5.4 Dirac Systems and Dirac Oscillators

We have so far considered systems related to a given hyperkahler structure. As we have discussed in Sect. 1.5, there is a notion of dual hyperkahler structures; in particular, in dimension four each hyperkahler structure has a dual one, and one could consider (integrable) systems related to a given *pair of dual* hyperkahler structures; this is the case e.g. for the hyperhamiltonian description of the Dirac equation (see the discussion in Sect. 7.2), and hence one speaks of *Dirac systems*.[7]

Definition 5.3 If the dynamical system $\dot{x} = f(x)$ in $\mathbf{R}^4$ is described by a vector field

$$
X = X_{(+)} + X_{(-)} \tag{5.36}
$$

which decompose as the sum of vector fields $X_{(\pm)}$ which are hyperhamiltonian with respect to a pair of dual (positively and negatively oriented) hyperkahler structures, we say it is a *Dirac system*, and X is a *Dirac vector field*. If $X_{(\pm)}$ both describe the flow of (simple) quaternionic oscillators, we are dealing with a (simple) *Dirac oscillator*.

---

[7]We recall also that in $\mathbf{R}^4$ any Dirac vector field is also strictly Dirac.

It will turn out that Dirac systems can be dealt with through Walcher factorization principle[8] [144].

This fact is at the origin of the seemingly restrictive definition of Dirac systems and Dirac oscillators for the higher dimensional case, i.e. for systems in $\mathbf{R}^{4n}$; this makes use of the notion of strictly Dirac vector fields, introduced in Sect. 2.4 and recalled here for convenience.

Definition 5.4 If the dynamical system $\dot{x} = f(x)$ in $\mathbf{R}^{4n}$ is described by a strictly Dirac vector field, i.e. by a vector field $X = X_{(+)} + X_{(-)}$ which decompose as the sum of vector fields $X_{(\pm)}$ which are hyperhamiltonian with respect to a pair of hyperkahler structures which:

1. are dual to each other (hence define the same partitioning of $\mathbf{R}^{4n}$ into irreducible $\mathbf{R}^4$ hyperkahler submanifolds)
2. define *in each* irreducible $\mathbf{R}^4$ hyperkahler submanifold the *opposite* orientation,

we say it is a *strictly Dirac system*, and X is a *strictly Dirac vector field*. If $X_{(\pm)}$ both describe the flow of quaternionic oscillators, we are dealing with a *Dirac oscillator*.

We will now consider this kind of situation, in fact for (general) Dirac oscillators. We stress they represent a generalization of quaternionic oscillators, the latter being obtained when only one of the vector fields $X_\pm$ is present (in each block); they have been introduced in [69].

We will work directly with the standard hyperkahler structures (usual considerations about the possibility of reducing to this case apply; note that as changes of coordinates should be orthogonal to preserve the metric, they do also preserve ρ); that is, we will have

$$X_{(\pm)} = f^i_{(\pm)}(x)\, \partial_i \,, \tag{5.37}$$

with coefficients $f^i_{(\pm)}$ given by

$$f^i_{(+)}(x) = \sum_{\alpha=1}^{3} c_\alpha(\rho)\, (Y_\alpha)^i{}_j\, x^j \,, \quad f^i_{(-)}(x) = \sum_{\alpha=1}^{3} \widehat{c}_\alpha(\rho)\, (\widehat{Y}_\alpha)^i{}_j\, x^j \,; \tag{5.38}$$

here $\rho = x_1^2 + \ldots + x_4^2 = |\mathbf{x}|^2$, and the matrices Y_α, $\widehat{Y}_\alpha$ are the ones defined in (1.23), (1.25).

We write

$$\nu_{(+)}(\rho) := \sqrt{c_1^2(\rho) + c_2^2(\rho) + c_3^2(\rho)} \,,$$
$$\nu_{(-)}(\rho) := \sqrt{\widehat{c}_1^2(\rho) + \widehat{c}_2^2(\rho) + \widehat{c}_3^2(\rho)} \,; \tag{5.39}$$

---

[8]As we will see below, this is effective only if the flows of $X_{(\pm)}$ can actually be integrated; thus in practice only for Dirac oscillators. The same will apply for higher dimensions.

and correspondingly

$$K_{(+)} = [1/\nu_{(+)}(\rho)] \sum_{\alpha=1}^{3} c_\alpha(\rho) (Y_\alpha)^i{}_j ;$$
$$K_{(-)} = [1/\nu_{(-)}(\rho)] \sum_{\alpha=1}^{3} \widehat{c}_\alpha(\rho) (\widehat{Y}_\alpha)^i{}_j . \tag{5.40}$$

We also write

$$K = K_{(+)} + K_{(-)} . \tag{5.41}$$

Note that

$$[K_{(+)}, K_{(-)}] = 0 ; \tag{5.42}$$

this follows at once from $[Y_\alpha, \widehat{Y}_\beta] = 0$ for any α, β.

The dynamics under X [and that under $X_{(\pm)}$] will then be described by, respectively,

$$\dot{x} = K x , \quad \left[\text{and } \dot{x} = K_{(\pm)} x \right] . \tag{5.43}$$

Each of the dynamics under $X_{(\pm)}$ corresponds to a quaternionic oscillator and can be integrated as shown in previous sections of this Chapter. We will now show that the dynamics under X is also explicitly integrable.

It follows from $Y_\alpha^T = -Y_\alpha$ and $\widehat{Y}_\alpha^T = -\widehat{Y}_\alpha$ that the matrices $K_{(\pm)}$ and K are also antisymmetric. This also implies that $\rho = |\mathbf{x}|^2$ is constant on the dynamics; thus once again we can consider the c_α and $\widehat{c}_\alpha$ coefficients (which depend on x only through ρ) as constant. The same holds for $\nu_{(\pm)}(\rho)$ (we will thus from now on just omit to indicate their dependence on ρ when dealing with a single solution).

Let us now look at the flow under X as in (5.37); Walcher's factorization principle [144] states that denoting by $\Phi(t; x_0; Y)$ the time t flow issuing from x_0 at time $t = 0$ under the vector field Y, under the condition $[X_{(+)}, X_{(-)}] = 0$ we have

$$\Phi(t; x_0; X) = \Phi\left[t; \Phi(t; x_0; X_{(+)}); X_{(-)}\right] = \Phi\left[t; \Phi(t; x_0; X_{(-)}); X_{(+)}\right] . \tag{5.44}$$

In the case we are presently considering, the flows under $X_{(\pm)}$ do indeed commute,[9] and actually can be explicitly computed, see Sect. 5.2.3; more precisely, with the present notation, we have

$$\Phi(t; x_0; X_{(\pm)}) = \left[\cos(\nu_{(\pm)}t) \mathcal{E}_4 + \sin(\nu_{(\pm)}t) K_{(\pm)}\right] x_0 := A_{(\pm)} x_0 . \tag{5.45}$$

It should also be noted that $[K_{(+)}, K_{(-)}] = 0$ entails $[A_{(+)}, A_{(-)}] = 0$ as well.

Remark 5.9 With the present notation, and writing for short $\chi_\pm = \cos(\nu_{(\pm)}t)$, $\sigma_\pm = \sin(\nu_{(\pm)}t)$, the matrices $A_{(\pm)}$ are given in explicit terms by

---

[9]This follows from taking both the $\mathcal{H}^\alpha$ *and* the $\widehat{\mathcal{H}}^\alpha$ to be functions of ρ alone.

$$A_{(+)} = \begin{pmatrix} \chi_+ & c_1\sigma_+ & c_3\sigma_+ & c_2\sigma_+ \\ -c_1\sigma_+ & \chi_+ & c_2\sigma_+ & -c_3\sigma_+ \\ -c_3\sigma_+ & -c_2\sigma_+ & \chi_+ & c_1\sigma_+ \\ -c_2\sigma_+ & c_3\sigma_+ & -c_1\sigma_+ & \chi_+ \end{pmatrix} ; \tag{5.46}$$

$$A_{(-)} = \begin{pmatrix} \chi_- & -\widehat{c}_3\sigma_- & \widehat{c}_1\sigma_- & -\widehat{c}_2\sigma_- \\ \widehat{c}_3\sigma_- & \chi_- & \widehat{c}_2\sigma_- & \widehat{c}_1\sigma_- \\ -\widehat{c}_1\sigma_- & -\widehat{c}_2\sigma_- & \chi_- & \widehat{c}_3\sigma_- \\ \widehat{c}_2\sigma_- & -\widehat{c}_1\sigma_- & -\widehat{c}_3\sigma_- & \chi_- \end{pmatrix} . \tag{5.47}$$

It is then immediate to check that in fact $[A_{(+)}, A_{(-)}] = 0$.   ☉

According to (5.44), the flow under X will then be described by

$$\Phi(t; \mathbf{x}_0; X) = A_{(+)}(t) A_{(-)}(t) \mathbf{x} = A_{(-)}(t) A_{(+)}(t) \mathbf{x} := A(t) \mathbf{x}; \tag{5.48}$$

this matrix $A(t)$ can also be written as

$$A(t) = [\cos(\nu_{(+)}t) \cos(\nu_{(-)}t)] I + [\cos(\nu_{(+)}t) \sin(\nu_{(-)}t)] K_{(+)} \tag{5.49}$$
$$+ [\sin(\nu_{(+)}t) \cos(\nu_{(-)}t)] K_{(-)} + [\sin(\nu_{(+)}t) \sin(\nu_{(-)}t)] K_{(+)} K_{(-)} .$$

A more explicit expression is immediately obtained by multiplying the two matrices $A_{(\pm)}$ as given in (5.46) and (5.47); this is long and not specially interesting and hence will not be reported here.

It is a matter of straightforward—albeit rather boring—algebra to check that indeed $\mathbf{x}(t) = A(t)\mathbf{x}_0$ is a solution to $d\mathbf{x}/dt = X(\mathbf{x}) = K\mathbf{x}$, for any initial condition $\mathbf{x}(0) = \mathbf{x}_0$; this is also seen by simply checking that $(dA/dt) = KA$.

Remark 5.10 It is quite remarkable that systems associated to quaternionic or Dirac oscillators via adding a gradient vector field present the phenomenon of *spontaneous linearization* [44], which in this context means *asymptotic integrability* [58].

In fact, let us consider a system of the type

$$\dot{x}^i = f_0(|\mathbf{x}|^2)x^i + \sum_\alpha c_\alpha(|\mathbf{x}|^2) (Y_\alpha)^i_{\ j} x^j + + \sum_\alpha \widehat{c}_\alpha(|\mathbf{x}|^2) (\widehat{Y}_\alpha)^i_{\ j} x^j ; \tag{5.50}$$

then the dynamics of $\rho = |\mathbf{x}|^2$ is controlled by f_0 alone (due again to skew-symmetry of the Y_α and $\widehat{Y}_\alpha$),

$$\frac{d\rho}{dt} = 2 f_0(|\mathbf{x}|^2) |\mathbf{x}|^2 . \tag{5.51}$$

Thus it will evolve towards the stable zeros of f_0 (those with $f'(\rho_0) < 0$). On spheres with such radius, which are reached asymptotically by the dynamics, the system will behave as a Dirac oscillator—or a quaternionic one if only the c_α or only the $\widehat{c}_\alpha$ are nonzero—and hence is integrable.   ☉

Remark 5.11 This discussion is immediately generalized to the case where the c_α depend not just on $\rho = |\mathbf{x}|^2$, but (with the notation introduced in Sect. 5.2) on the $\rho_k = |\xi_{(k)}|^2$ as well. That is, systems of the form

$$\dot{x}^i = f_0(\rho_1, \ldots, \rho_n)x^i + \sum_\alpha c_\alpha(\rho_1, \ldots, \rho_n)(Y_\alpha)^i_{\ j}x^j$$
$$+ \sum_\alpha \widehat{c}_\alpha(\rho_1, \ldots \rho_n)(\widehat{Y}_\alpha)^i_{\ j}x^j .$$

In fact, the skew-symmetry of the Y_α and $\widehat{Y}_\alpha$ again makes that the dynamics of the ρ_k only depends on f_0; more precisely, we have

$$\frac{d\rho_k}{dt} = 2 f_0(\rho_1, \ldots, \rho_n) \rho_k . \tag{5.52}$$

Thus in this case the "radial" (in each block) degrees of freedom evolve autonomously; however we have an n-dimensional (reduced) dynamical system, and its dynamics can be rather complex. If and when it drives the ρ_k towards a fixed point, we have again spontaneous linearization and asymptotic integrability; otherwise we can formally write the evolution of the angular, i.e. "spin", degrees of freedom as a linear but non autonomous system, driven by the (generally, non integrable[10]) evolution of the radial, i.e. "action" variables.   ⊙

5.5 Symmetry Versus Integrability for Hyperhamiltonian Systems

In the standard Hamiltonian case, integrability and symmetry (under an abelian group of suitable dimension) are intimately related. In this Section we will discuss the interrelations between symmetry and integrability for hyperhamiltonian systems. Our discussion follows the one provided in [59, 69].

5.5.1 Symmetry of Integrable Hyperhamiltonian Systems

We do now want to discuss the symmetry properties of integrable hyperhamiltonian systems; here we will consider the special hyperhamiltonian systems which are Hamiltonian as degenerate systems, and focus instead on the generic case of non-Hamiltonian integrable hyperhamiltonian systems (see [59] for a discussion).[11]

---

[10]This is why we say that writing the evolution of angular degrees of freedom in linear form is just formal.

[11]It is maybe worth recalling what is precisely meant when we speak of symmetry for a dynamical system described by a vector field X on a smooth manifold M. We consider smooth maps $\Phi : M \to M$, and their push-forward $\Phi_* : TM \to TM$; we say that Φ is a symmetry of (the dynamical system characterized by) X if $\Phi_*(X) = X$. One is often interested in one-parameter (or multi-parameter) families of maps $\Phi_\alpha : M \to M$, generated by a vector field $Z : M \to TM$; in this case one says,

As quaternionic integrable systems are characterized by the property of being diffeomorphic to a system of quaternionic oscillators – and since symmetry properties are conserved under diffeomorphisms – we can investigate the symmetry properties of quaternionic integrable systems by working directly on (5.10), or equivalently on (5.12).

We start by considering the simple quaternionic oscillator in $\mathbf{R}^4$, see (5.7), and want to determine its symmetries; we assume that the system is not Hamiltonian, which implies that (at least some of) the c_α do actually depend on the ξ_α, i.e. that $L(\rho)$ is not constant.

We will just consider *linear* transformations; this is not really a limitation. In fact, if we denote by $\mathcal{G}_0$ the Lie algebra of linear symmetries, and by $\mathcal{G}$ that of general Lie-point symmetries, it turns out that

$$\mathcal{G} = \mathcal{I} \otimes \mathcal{G}_0 , \tag{5.53}$$

where $\mathcal{I}$ is the ring of smooth constants of motions for X, i.e. in this case the ring of smooth functions of ρ. This actually holds for general dynamical systems, see [46].

It should be stressed that here we have to make a choice: that is, we can consider the equations of motion as a generic dynamical system and look for its symmetries irrespective of the preservation (under these maps) of the metric and of the hyper-kahler structure; or we could be looking only for symmetries which preserve these structures, i.e. which are *hyperkahler maps* [68].

Thus we will consider maps (sum over repeated latin indices is understood):

$$\widetilde{x}^i = \Lambda^i{}_j x^j , \tag{5.54}$$

where Λ is a constant regular matrix. As discussed in Sect. 5.1, the trajectories lie on spheres with constant radius; it follows that symmetries should preserve ρ as well, i.e. that the matrix Λ should be orthogonal, $\Lambda \in O(4)$[12]; this also agrees, of course, with the requirement of preserving the metric in M. In fact, if we want to preserve orientation (if not, a hyperkahler structure would be mapped into its dual; see above) we will conside $\Lambda \in SO(4)$.

Substituting this map in the equations of motion we get

$$\frac{d\widetilde{x}^i}{dt} = \Lambda^i{}_j \dot{x}^j = \sum_\alpha c_\alpha \, \Lambda^i{}_j \, (L_\alpha)^j{}_k \, (\Lambda^{-1})^k{}_\ell \, \widetilde{x}^\ell ; \tag{5.55}$$

thus the system is invariant iff, for all $i = 1, \ldots, 4n$,

---

(Footnote 11 continued)
with a slight abuse of language, that Z is a symmetry (or a local symmetry) of X if the one-parameter group Φ_α generated by Z is a group (or a local group) of symmetries of X.

[12]This follows from the requirement that the c_α and hence the matrix L do effectively depend on ρ (see above); should these be actually constant, a dilation would (commute with time evolution and hence) be a symmetry [32].

$$\sum_\alpha c_\alpha \, \Lambda^i{}_j \, (L_\alpha)^j{}_k \, (\Lambda^{-1})^k{}_\ell = \sum_\alpha c_\alpha \, (L_\alpha)^i{}_\ell \, . \qquad (5.56)$$

A sufficient condition for this to hold is

$$\Lambda \, L_\alpha \, \Lambda^{-1} = L_\alpha \quad (\alpha = 1, 2, 3) \, ; \qquad (5.57)$$

maps satisfying this conditions can be called *strong symmetries*.

Remark 5.12 The extension to any $\mathbf{R}^{4n}$ space is straightforward, once the matrices L_α have been block-diagonalized. We have to determine the group of linear transformations $\Lambda \in SO(4n)$ satisfying

$$\Lambda \, L_\alpha \, \Lambda^{-1} = L_\alpha \, , \quad (\alpha = 1, 2, 3) \, . \qquad (5.58)$$

As the L_α satisfy the quaternionic relations, they generate a linear representation (reducible into the sum of representations of dimension 4) of the Lie algebra su(2); we can thus work with four-dimensional matrices, and (5.57) becomes (with an obvious abuse of notation)

$$\Lambda \, L_\alpha = L_\alpha \, \Lambda \, . \qquad (5.59)$$

The representation is real irreducible, but complex reducible. Thus the (real) Schur lemma (see e.g., Chap. 8 of [105]) implies that Λ is an element of the subgroup of $SO(4)$ generated by the matrices of the structure dual to that generated by the matrices L_α; this is just SO(3). Recall that in fact SO(4) $\sim$ SO(3) $\times$ SO(3). ⊙

Equation (5.57) guarantees the invariance of the system, but as mentioned above we can also consider the more general condition (which obviously contains (5.57) as a particular case) given in Eq. (5.56).

Remark 5.13 In order for this to be satisfied, it is necessary that the matrices L_α transformed (by conjugation) under Λ are a linear combination of the original ones,

$$\widetilde{L}_\alpha = \Lambda \, L_\alpha \, \Lambda^{-1} = \sum_\beta R_{\alpha\beta} \, L_\beta \, . \qquad (5.60)$$

As easy to prove, if we moreover require that the $\widetilde{L}_\alpha$ satisfy the quaternionic relations, then the matrix R is necessarily a rotation in $\mathbf{R}^3$ and the invariance condition is

$$\sum_\beta R_{\alpha\beta} \, c_\beta = c_\alpha \, , \qquad (5.61)$$

which implies that the three-dimensional vector $\mathbf{c} = (c_1, c_2, c_3)$ is an eigenvector of the rotation R (and of its inverse); equivalently, the matrix R is a rotation with a fixed axis (identified by $\mathbf{c}$), and hence the group generated by the admitted R is SO(2). ⊙

In order to determine the possible matrices Λ satisfying Eq. (5.56), it is convenient to work at the infinitesimal level. At first order in ε, we have

$$\Lambda = I_{4n} + \varepsilon \lambda, \quad \lambda + \lambda^T = 0; \quad R = I_3 + \varepsilon r, \quad r + r^T = 0. \qquad (5.62)$$

Here λ are antisymmetric matrices, generating the group of admissible Λ and to be determined; and the matrix r is the generator of the SO(2) group of rotations with axis $\mathbf{c}$, i.e.

$$\sum_{\beta} r_{\alpha\beta} c_{\beta} = 0. \qquad (5.63)$$

With these, the invariance equation (5.60) is transformed into the *infinitesimal invariance equation*

$$[\lambda, L_\alpha] = \sum_{\beta=1}^{3} r_{\alpha\beta} L_\beta, \quad (\alpha = 1, 2, 3). \qquad (5.64)$$

Solving this is not too difficult in the case we are considering, i.e. for Euclidean $\mathbf{R}^{4n}$ spaces. It will be convenient to work first for $n = 1$ and then consider the case of higher n; actually the case of general n will be essentially the same as for $n = 2$.

5.5.1.1 The Case $n = 1$

For $n = 1$ any skew-symmetric matrix $\lambda \in so(4)$ can be written as a linear combination of the quaternionic matrices Y_α and $\widehat{Y}_\alpha$, corresponding to the two su(2) algebras in the decomposition

$$so(4) = su(2) \oplus su(2) = sp(1) \oplus sp(1);$$

recall that these satisfy $[Y_\alpha, \widehat{Y}_\beta] = 0$ for all α, β.[13]
Thus we set

$$\lambda = \frac{1}{2} \sum_{\beta=1}^{3} a_\beta Y_\beta + \frac{1}{2} \sum_{\beta=1}^{3} \widehat{a}_\beta \widehat{Y}_\beta, \qquad (5.65)$$

with $a_\beta, \widehat{a}_\beta$ undetermined coefficients. Let us consider, for definiteness, the positively oriented standard structure Y_α. Then the infinitesimal invariance equation (5.64) yields the solution

---

[13] Here and in the following sp(n) is the Lie algebra of the group Sp(n) of unitary four-dimensional symplectic matrices. To avoid any misunderstanding (and in view of some ambiguity in the literature for the notation for symplectic groups), we state explicitly that here the group Sp(n) is the set of $2n \times 2n$ (complex) unitary symplectic matrices – thus with real representation of dimension $4n$— with Lie algebra sp(n) $\subset Mat(2n; \mathbf{C}) \simeq Mat(4n; \mathbf{R})$.

$$r_{\alpha\beta} = \sum_{\gamma=1}^{3} \epsilon_{\alpha\beta\gamma} a_{\gamma} . \tag{5.66}$$

As r satisfies Eq. (5.63), the vector $\mathbf{a}$ should satisfy

$$\sum_{\gamma} \epsilon_{\alpha\beta\gamma} c_{\beta} a_{\gamma} = 0; \tag{5.67}$$

this implies that $\mathbf{a}$ is proportional to $\mathbf{c}$, and

$$r_{\alpha\beta} = \sum_{\gamma} \epsilon_{\alpha\beta\gamma} c_{\gamma} . \tag{5.68}$$

In conclusion, we have shown that:

Lemma 5.1 *The invariance algebra for the quaternionic oscillator equations* (5.7) *in dimension 4 is*

$$\mathcal{L}_1 = so(2) \oplus su(2) \approx so(2) \oplus sp(1) , \tag{5.69}$$

and the generator of the subalgebra $so(2)$ *is determined by the constants* c_{α}, *see* (5.68).

5.5.1.2 The Case $n = 2$

Let us now consider $n = 2$, i.e. $M = \mathbf{R}^8$. The matrices L_α are a hyperkahler structure, and we will assume that they are written as 4×4 diagonal block matrices (the invariance group for an arbitrary hyperkahler structure of the same orientation will be conjugated to the ones obtained in this way), each block having a definite orientation. It is easy to show that all of these structures are conjugated under $O(8)$ and then we can reduce the study to that of a positive oriented hyperkahler structure, i.e.

$$L_\alpha = \begin{pmatrix} Y_\alpha & 0 \\ 0 & Y_\alpha \end{pmatrix} \tag{5.70}$$

Lemma 5.2 *The symmetry algebra of the quaternionic oscillator in* $\mathbf{R}^8$ *is*

$$\mathcal{L}_2 = so(2) \oplus sp(2) . \tag{5.71}$$

Proof The proof goes through some explicit computations, which we will sketch here (see [70] for details). We build a basis of $o(8)$ using the Y_α and $\widehat{Y}_\alpha$ as building blocks; we will actually also need some symmetric 4×4 matrices. The basis (made of 28 matrices) can be written as

$$\begin{pmatrix} Y_\alpha & 0 \\ 0 & 0 \end{pmatrix}, \begin{pmatrix} 0 & 0 \\ 0 & Y_\alpha \end{pmatrix}, \begin{pmatrix} \widehat{Y}_\alpha & 0 \\ 0 & 0 \end{pmatrix}, \begin{pmatrix} 0 & 0 \\ 0 & \widehat{Y}_\alpha \end{pmatrix}$$

$$\begin{pmatrix} 0 & Y_\alpha \\ Y_\alpha & 0 \end{pmatrix}, \begin{pmatrix} 0 & \widehat{Y}_\alpha \\ \widehat{Y}_\alpha & 0 \end{pmatrix}, \begin{pmatrix} 0 & S_i \\ -S_i & 0 \end{pmatrix}. \tag{5.72}$$

with $\alpha = 1, 2, 3$, and S_i, $i = 1, \ldots, 10$, the set of 4×4 elementary symmetric matrices (that is, E_{jj} and $E_{jk} + E_{kj}$, where E_{jk} is the elementary matrix with 1 in the position jk and 0 elsewhere).

Applying the Eq. (5.64), we get the ten matrices λ

$$\begin{pmatrix} \widehat{Y}_\alpha & 0 \\ 0 & 0 \end{pmatrix}, \begin{pmatrix} 0 & 0 \\ 0 & \widehat{Y}_\alpha \end{pmatrix}, \begin{pmatrix} 0 & \widehat{Y}_\alpha \\ \widehat{Y}_\alpha & 0 \end{pmatrix} \begin{pmatrix} 0 & I_4 \\ -I_4 & 0 \end{pmatrix}. \tag{5.73}$$

The matrix r is

$$r = \begin{pmatrix} r_0 & 0 \\ 0 & r_0 \end{pmatrix}, \tag{5.74}$$

where r_0 is the matrix we found in the $n = 1$ (four-dimensional) case; it generates the algebra so(2). The matrices λ generate a Lie algebra of dimension ten which commutes with the above algebra so(2), and leaves invariant each of the L_α.

These matrices turn out to be a representation of the symplectic algebra sp(2), hence we have $\mathcal{L}_2 = $ so(2) $\oplus$ sp(2), as claimed.   △

5.5.1.3 The Case $n > 2$

The general case, i.e. quaternionic oscillators in $\mathbf{R}^{4n}$, is a generalization of the $n = 2$ (eight-dimensional case) and can be dealt with through similar computations, which we do not give here (see [70] for details); we get

Lemma 5.3 *The symmetry algebra of the quaternionic oscillator in* $\mathbf{R}^{4n}$ *is*

$$\mathcal{L}_n = \text{so}(2) \oplus \text{sp(n)} . \tag{5.75}$$

Remark 5.14 This fact is closely related to the computation of the holonomy group of hyperkahler and quaternionic manifolds, see for instance [21, 100, 133]; but we will not discuss this relation here.   ⊙

Remark 5.15 Finally, we note that su(2) $\oplus \ldots \oplus$ su(2) $\subset$ sp(n); this means that the symmetry algebra $\mathcal{L}_n$ include the product of n independent su(2) algebras, each acting on one (quaternionic) degree of freedom.   ⊙

5.5.2 Integrable Hyperhamiltonian Systems and Symmetry

We have so far discussed and characterized the symmetry of a quaternionic integrable system. Now we will reverse our point of view, and discuss systems which enjoy the same symmetry properties of quaternionic integrable system.

The motivation for this discussion is as follows. In the Hamiltonian case (with compact energy manifolds), it is well known that a torus symmetry is enough to conclude that the system is integrable (see e.g. [10]); we want to investigate if something similar holds in the quaternionic case.

Remark 5.16 In the Hamiltonian case (with compact energy manifolds), the symmetry is sufficient to fully characterize integrable systems thanks to a "topological lemma" (see e.g. Lemma 2 of sect. 49, p. 274, in [10]), which guarantees that *if a compact connected n-dimensional manifold admits an abelian algebra of n tangent vector fields spanning a regular n-dimensional distribution, then it is a torus* $\mathbf{T}^n$. We are not able to provide a similar statement for the quaternionic case, for the lack of a similar result for manifolds admitting an algebra of tangent vector fields corresponding to the $\mathcal{L}_n$ algebra. (See also Remark 5.18.)   ⊙

First of all, we note that quaternionic oscillators in n degrees of freedom, i.e. in $\mathbf{R}^{4n}$, admit as invariant manifolds the product

$$V^n := S^3 \times \ldots \times S^3 \quad (n \text{ factors}) , \qquad (5.76)$$

where each S^3 factor belongs to an invariant $\mathbf{R}^4$ subspace; admitting such an invariant manifold is of course a *necessary* condition for a system to be a quaternionic oscillator.

It will again be convenient to treat separately the $n = 1$ case and that for general $n > 1$.

5.5.2.1 The Case $n = 1$

Let us now consider the $\mathcal{G} = su(2)$ algebra spanned by the $\widehat{Y}_\alpha$ matrices, and look for vector fields $X = f^i \partial_i$ in $\mathbf{R}^4$ which are symmetric under this and do moreover leave the spheres S^3 invariant (i.e. admit ρ as constant of motion; equivalently, preserve the Euclidean metric); it is easy (e.g. by explicit computation) to check that these vector fields reduce to the module over function of ρ (i.e. to linear combinations with coefficients depending on ρ) generated by those associated to the Y_α.[14] In other words, we have

$$X = \left[\sum_\alpha c_\alpha(\rho) \, (Y_\alpha)^i{}_j x^j \right] \partial_i . \qquad (5.77)$$

---

[14]The requirement to leave spheres invariant does of course exclude the dilation vector fields associated to the identity matrix.

Conversely, if we look at the $\mathcal{G} = su(2)$ algebra spanned by the Y_α matrices, and look for vector fields $\widehat{X} = f^i \partial_i$ in $\mathbf{R}^4$ which are symmetric under this—and do moreover leave the spheres S^3 invariant—it results

$$\widehat{X} = \left[\sum_\alpha \widehat{c}_\alpha(\rho) \, (\widehat{Y}_\alpha)^i{}_j x^j \right] \partial_i \, . \tag{5.78}$$

Remark 5.17 Albeit this result is easily checked by explicit computation, its true nature follows from the Schur lemma in its real version [105]: the only matrices which commute with the whole irreducible representation of $\mathcal{G}$ spanned by the $\widehat{Y}_\alpha$ (respectively, by the Y_α) matrices, are the identity and those of the conjugated representation, i.e. the Y_α (respectively, the $\widehat{Y}_\alpha$).[15]

This is then rephrased in terms of vector fields; more precisely, the vector fields must be expressed in this way at each point. One could then think of coefficients c_α (respectively $\widehat{c}_\alpha$) being functions of $(x_1, \ldots, x_4)$, but plugging such a vector field into the symmetry condition yields that they can actually depend on the x_i only through $\rho = |x|^2$.

To make the argument completely clear, just consider the case where $\mathcal{G}$ is spanned by the $\widehat{Y}_\beta$ and denote by $\mathcal{Y}_\alpha$ the vector field $\mathcal{Y}_\alpha = (Y_\alpha)^i{}_j x^j \partial_i$ associated to the matrix Y_α, and similarly for $\widehat{\mathcal{Y}}_\alpha$. The dynamical vector field is, as follows from the Schur Lemma argument, $X = \sum_\alpha c_\alpha(x) \mathcal{Y}_\alpha$; then we immediately have that $[\widehat{\mathcal{Y}}_\beta, X] = \sum_\alpha \widehat{\mathcal{Y}}_\beta(c_\alpha) \, \mathcal{Y}_\alpha$. As the Y_α (and hence the $\mathcal{Y}_\alpha$) are independent, this can vanish only if $\widehat{\mathcal{Y}}_\beta(c_\alpha) = 0$ for all α and β, i.e. if the c_α only depends on joint invariants for the $\widehat{\mathcal{Y}}_\beta$; but the only function which is invariant under the three vector fields $\widehat{\mathcal{Y}}_\beta$ is $\rho = |x|^2$. The same holds, obviously, if we interchange the $\widehat{\mathcal{Y}}_\alpha$ and the $\mathcal{Y}_\alpha$.   ⊙

Note that on each sphere S^3 the system will automatically have an SO(2) symmetry, corresponding to rotations around an axis identified by the $c_\alpha(\rho)$.

5.5.2.2   The Case of Arbitrary $n > 1$

The general case (arbitrary n) is discussed along the same lines. We recall

$$\mathcal{G} := su(2) \oplus \ldots \oplus su(2) \subset sp(n) \, ,$$

and start considering the subalgebra $\mathcal{G} \subset \mathcal{L}_n$ acting in $\mathbf{R}^{4n} = \mathbf{R}^4 \oplus \ldots \oplus \mathbf{R}^4$ by leaving each $\mathbf{R}^4$ subspace invariant. Note it also preserves the $V_n = S^3 \times \ldots \times S^3$ manifolds, where each S^3 belongs to one of the $\mathbf{R}^4$ in the splitting mentioned above. The action of $\mathcal{G}$ in each $\mathbf{R}^4_{(m)}$ $(m = 1, \ldots, n)$ invariant subspace is described by a

---

[15]As mentioned in the previous footnote, in our case the identity is ruled out by the requirement to preserve ρ, or equivalently the Euclidean metric.

linear combination of either the Y_α or the $\widehat{Y}_\alpha$ matrices; without any loss of generality, we can take as generators exactly either[16] $L_\alpha^{(m)} = Y_\alpha$ or $L_\alpha^{(m)} = \widehat{Y}_\alpha$.

We thus look for vector fields in $\mathbf{R}^{4n}$ which are symmetric under $\mathcal{G}$. Proceeding as in the $n = 1$ case, we first deal with matrices; writing this in four-dimensional block form, one easily concludes that elements on the block diagonal must commute with the corresponding $L_\alpha^{(m)}$ matrices, while those in off-diagonal blocks vanish (again this follows from the real version of Schur Lemma [105]).

Thus in each block we are left with matrices in the conjugated representation of the su(2) Lie algebra. In other words, at each point we have a vector field associated to the conjugated representation of SU(2) $\times \ldots \times$ SU(2); when we pass to vector fields on $\mathbf{R}^{4n}$ we will have linear combinations of these with coefficients which could in principles depend on the coordinates, but (as in the case $n = 1$) when we require the commutation with the symmetry vector fields and invariance of the $V_n = S^3 \times \ldots \times S^3$ manifolds, we are only left with the possibility of coefficients depending on $|\xi_1|^2, \ldots, |\xi_n|^2$, with the same argument seen above in the $\mathbf{R}^4$ case.

We summarize our discussion as follows. We consider $\mathbf{R}^{4n}$ and the group $G = $ SU(2) $\times \ldots \times$ SU(2), each factor acting effectively on a subspace $\mathbf{R}^4$ of $\mathbf{R}^{4n}$ and trivially on the other ones, and with the action of G in the m-th subspace $\mathbf{R}^4_{(m)}$ described by the SU(2) action generated by the Y_α matrices (respectively by the $\widehat{Y}_\alpha$ ones). The vector fields in $\mathbf{R}^{4n}$ which are symmetric under this group action are precisely those corresponding to quaternionic oscillators with suitable signature.

These possess in turn additional symmetry as described by $\mathcal{L}_n$ identified in the previous Section. In other words, *in this framework the symmetry properties characterize quaternionic integrable systems.*

Remark 5.18 It should be stressed that we do *not* have an analogue of the result holding for Hamiltonian integrable systems. In fact, here we had to explicitly *require* invariance of the V_n manifolds; on the other hand, in the Hamiltonian case the symmetry properties were sufficient to characterize the topology of the invariant manifolds (i.e. tori $\mathbf{T}^n$) and integrability followed from this.

The point is that, to the best of our knowledge, one cannot state any correspondence between invariance under SU(2) $\times \ldots \times$ SU(2) and the topology of the manifolds $V_n = S^3 \times \ldots \times S^3$, contrary to what happens for U(1) $\times \ldots \times$ U(1) and $\mathbf{T}^n$.

This entails in particular that our discussion—conducted in the Euclidean framework—cannot be extended to the general case as we are not able to control the global geometry in the general setting; in other words, we have a result which only holds locally on each chart.   ⊙

---

[16]For each $m = 1, \ldots, n$ we have either Y or $\widehat{Y}$, but the same option must be chosen for all $\alpha = 1, 2, 3$.

Chapter 6
Hyperhamiltonian Dynamics and Perturbation Theory

In the previous chapter we have studied integrable hyperhamiltonian systems. We want now to study hyperhamiltonian *perturbations* of integrable systems. By this, we mean in general hyperhamiltonian perturbations of hyperhamiltonian integrable systems; however, as we have seen, standard Hamiltonian systems are a special case of hyperhamiltonian systems, and the same holds for integrable systems.

Here we will consider perturbation theory in its fundamental form, dating back to Poincaré, i.e. through reduction of a system to *normal form* near an equilibrium point.

We start by recalling the classical normalization procedure for vector fields and for Hamiltonian systems. Then we formulate the theory for hyperhamiltonian systems.

6.1 Perturbation Theory and Normal Forms

We will collect here some basic and well known facts about perturbation theory for dynamical systems in general, and for Hamiltonian systems in particular. A detailed discussion is provided e.g. in [7, 10–12, 55, 97]; we will follow the notation used in [46].

6.1.1 General Dynamical Systems

We will always work in the neighborhood $U \subseteq M$ of a fixed point m_0; thus we can use local coordinates x^i in U (in these the fixed point will just be the origin of the coordinate system).

© Springer International Publishing AG 2017
G. Gaeta and M.A. Rodríguez, *Lectures on Hyperhamiltonian
Dynamics and Physical Applications*, Mathematical Physics Studies,
DOI 10.1007/978-3-319-54358-1_6

Let us consider a (analytic) vector field X written in coordinates as

$$X = f^i(x)\, \partial_i \; ; \tag{6.1}$$

by assumption $f^i(0) = 0$. This corresponds to the dynamical system

$$\dot{x}^i = f^i(x) \, . \tag{6.2}$$

We will expand the f^i in a Taylor series around $m_0 = 0$, and write

$$f^i(x) = \sum_{k=0}^{\infty} f_k^i(x) \; ; \tag{6.3}$$

here the f_k^i are terms homogeneous of order $(k + 1)$ in the x^i (the reason for this notation, i.e. for the shift in the order of f_k, will be clear in a moment).

In practice we will always use truncated series, i.e. write

$$f^i(x) = \sum_{k=0}^{N} f_k^i(x) + \mathcal{R}^i(x) \, , \tag{6.4}$$

where $\mathcal{R}^i$ is the remainder of the series.

If we are in the vicinity of the origin, i.e. for $|\mathbf{x}| \approx \varepsilon$, then higher order terms can be considered as perturbations of the lower order (linear) ones. In fact, the associated dynamical system (6.2) reads now

$$\dot{x}^i = f_0^i(x) + \sum_{k} \varepsilon^k f_k^i(x) \, . \tag{6.5}$$

We would of course be able to integrate the unperturbed system (that is, the linear one obtained for $\varepsilon = 0$). If we were able to determine a change of coordinates reducing the full system to its linear part (i.e., to *linearize* the system), the same would apply to the full system.

The procedure of *normalization* was devised by Poincaré as a tool to integrate nonlinear systems (at least near a fixed point). This is indeed possible if certain assumptions are verified, while in general one can only reduce the system to a simpler form—or more precisely to a form possessing certain convenient properties[1]—which goes under the name of Poincaré-Dulac *normal form*.[2]

---

[1] The notion of "simpler" is ambiguous, and actually what in the spirit of Poincaré theory is a simpler form can—and in general will—contain infinitely many terms, albeit one started from a very simple nonlinear expression.

[2] The theory was developed by Poincaré for non-resonant systems—in which case one can always arrive at least formally at a linear system—and by his pupil Dulac in the resonant case.

6.1.1.1 Poincaré Transformations

We will consider a *near-identity* change of coordinates; by this we mean a change of coordinates $x \to y$ of the form

$$y = x + h(x),$$

where h is at least quadratic in x; note this is near-identity—that is, $|h(x)| \ll |x|$—only for $|x| \approx 0$ (and anyway $|x| \ll 1$), so that when we speak of "near-identity" changes of coordinates we are implicitly assuming this condition. In our framework this is not a problem, as we are indeed actually working in a neighborhood U of the origin.

With this change of coordinates, the dynamical system (6.2) is mapped into

$$\dot{y}^i = \dot{x}^i + \left(\frac{\partial h^i}{\partial x^j}\right) \dot{x}^j = f[x + h(x)]. \tag{6.6}$$

We can define the matrix

$$\Gamma^i{}_j = \left(\frac{\partial h^i}{\partial x^j}\right),$$

so that (6.6) is written as

$$(I + \Gamma)^i{}_j \, \dot{x}^j = f^i[x + h(x)]; \tag{6.7}$$

as we assumed $|h(x)| \ll |x|$, in particular that h is at least quadratic in x, we also have $|\Gamma| \ll 1$, and

$$\Lambda := (I + \Gamma)^{-1}$$

is well defined near the origin. We can thus apply Λ on both sides of (6.7) and get

$$\dot{x}^i = \Lambda^i{}_j \, f^j[x + h(x)] := \tilde{f}^i(x). \tag{6.8}$$

This general discussion becomes more relevant—in particular, in that one can effectively determine the explicit expression for the "new" functions $\tilde{f}^i(x)$—when combined with the series expansion (6.3) for f *and* when one chooses a function $h(x)$ which is homogeneous of degree $(m+1)$ in x (with $m \geq 1$, see above); we will write h as $h_m(x)$ to keep note of its degree of homogeneity. In this case, one speaks of a *Poincaré transformation* (or Poincaré map).

Remark 6.1 In the modern formulation it is convenient to consider changes of coordinates generated by a *Lie transform* [20, 23, 24], see Sect. 6.1.3. One of the advantages of this is that certain inverse operations to be considered, see below, are automatically guaranteed to exist thanks to the properties of Lie groups. We will follow this approach, see Sects. 6.2.2 and 6.3. ⊙

6.1.2 Poincaré-Dulac Normal Forms

In order to implement this transformation, we need to express Λ and $f^j(x+h)$ in a series expansion; it will suffice to keep track of the first order corrections.

First of all, we note that

$$f_k^i[x + h_m(x)] = f_k^i(x) + \left(\frac{\partial f_k^i}{\partial x^j}\right) h_m^j(x) + \left(\frac{\partial^2 f_k^i}{\partial x^j \partial x^\ell}\right) h_m^j(x) h_m^\ell(x) + \cdots;$$

our notation makes it very easy to keep track of the degree of the different terms: the first one (no derivatives) is of degree $(k+1)$, the second one (first derivatives) is of degree $[((k+1)-1)+(m+1)] = (k+m+1)$, the third one (second derivatives) is of degree $[((k+1)-2)+2(m+1)] = (k+2m+1)$. It is then clear that subsequent terms in the Taylor expansion would be of order $(k + Nm + 1)$. We will thus write

$$f_k^i[x + h_m(x)] = f_k^i(x) + \varphi_{(k,m)}^i + \text{h.o.t.}. \tag{6.9}$$

Here the term $\varphi_{(k,m)}^i$ is homogeneous of degree $((k+m)+1)$. We stress that all the "correction" terms are at least of the same order as h_m (in fact, all terms in f are at least linear, i.e. $k \geq 0$).

On the other hand, recalling that $|\Gamma| \ll 1$, we have

$$\Lambda = (I + \Gamma)^{-1} = I - \Gamma + \frac{1}{2}\Gamma^2 + \cdots;$$

as we have assumed h to be homogeneous of degree $m + 1$, Γ is homogeneous of degree m, Γ^2 of degree $2m$, etc.; we thus write

$$\Lambda = I - \Gamma + \text{h.o.t.}. \tag{6.10}$$

We can now combine (6.9) and (6.10); this yields (neglecting h.o.t.)

$$\widetilde{f} = (I - \Gamma_{m-1}) \sum_k \left[f_k + \varphi_{(k,m)}\right]$$

$$= \sum_k \left[f_k(x) + \varphi_{(k,m)} - \Gamma f_k\right] + \text{h.o.t.}.$$

We can rewrite this as a series expansion,

$$\widetilde{f} = \sum_k \widetilde{f}_k; \tag{6.11}$$

the expression for the terms $\widetilde{f}_k$ are easily determined, at least at orders not higher than $m + 1$: in fact (recalling once again that Γ is homogeneous of degree m), we have

$$\widetilde{f}_k(x) = f_k(x) \quad (k < m) ; \tag{6.12}$$

$$\widetilde{f}_m(x) = f_m(x) + \varphi_{(0,m)} - \Gamma f_0 . \tag{6.13}$$

An explicit expression for $\widetilde{f}_k$ with $k > m$ could also be determined (see e.g. [46]), but it is not of interest here.

Our result is therefore that under a Poincaré map of degree $m + 1$, all terms of lower degree are unchanged, see (6.12); while those of the same degree change via the action of a linear operator $\mathcal{L}$, i.e.

$$f_m \rightarrow \widetilde{f}_m = f_m - \mathcal{L}(h_m) . \tag{6.14}$$

Recalling the expression of $\varphi_{(k,m)}$ and comparing with (6.13) we readily obtain

$$\mathcal{L}(h^i) := f_0^j \left(\frac{\partial h^i}{\partial x^j} \right) - \left(\frac{\partial f_k^i}{\partial x^j} \right) h^j . \tag{6.15}$$

If we recall that f_0 represents linear terms, we realize that it is always possible to write

$$f_0^i(x) = A^i{}_j x^j ;$$

in this way we have

$$\mathcal{L}(h^i) := \left(\frac{\partial h^i}{\partial x^j} \right) A^j{}_k x^k - A^i{}_j h^j . \tag{6.16}$$

The operator $\mathcal{L}$ is also known as the *homological operator*; note it maps homogeneous (vector) functions into homogeneous (vector) functions of the same degree. It is sometimes convenient—also in order to deal with operators acting in finite dimensional spaces—to consider the restrictions $\mathcal{L}_k$ of $\mathcal{L}$ to vector functions homogeneous of degree $(k + 1)$.

It is quite clear that, by a suitable choice of the generating function h_k, we can eliminate from f_k all terms which are in the range of $\mathcal{L}_k$; we refer to this as *normalization* at order $(k + 1)$. Moreover, thanks to (6.12), we can operate a sequence of Poincaré transformations with generating functions of increasing order, and at each step leaving unaffected the terms which have already been normalized at previous steps.

In other words, by an ordered sequence of Poincaré transformations it is always possible to transform a dynamical system into one which has only terms in the spaces complementary to the range of $\mathcal{L}_k$, up to any desired order N. We say then that the system is in *Poincaré-Dulac normal form* up to order N.

In order to do this we should choose h_m as being a solution to the *homological equation*

$$\mathcal{L}_m(h_m) = \pi_\rho(f_m) , \tag{6.17}$$

where π_ρ is the projection to the range of $\mathcal{L}$. Note that here f_m represents the function f_m after the previous steps of normalization (with generators $h_1, \ldots, h_{m-1}$) have been carried out.

Denoting by $\mathcal{L}_m^*$ the pseudo-inverse of $\mathcal{L}_m$, this yields

$$h_m = \mathcal{L}_m^*[\pi_\rho(f_m)] . \tag{6.18}$$

6.1.2.1 Description in Eigen-Coordinates

If the matrix A identifying the linear part of f is given explicitly, it is convenient to choose coordinates which give a basis of eigenvectors for (the semisimple part of) A; note that we will be interested in cases where A is semisimple, so we do not need to discuss several subtleties which are associated to the presence of a nilpotent part in the Jordan decomposition of a general A [80].

In these coordinates, A is diagonal; we will write it as

$$A = \text{diag}(\lambda_1, \ldots, \lambda_n) . \tag{6.19}$$

Now (6.15) simplifies to

$$\mathcal{L}(h^i) := \lambda_j x^j \left(\frac{\partial h^i}{\partial x^j}\right) - A^i{}_j h^j . \tag{6.20}$$

We will then write h_m as

$$h_m^i = c^i_{j_1 \ldots j_n} (x^1)^{j_1} \ldots (x^n)^{j_n} = c^i_J x^J , \tag{6.21}$$

where $c^i_{j_1 \ldots j_n}$ are real constants, j_i are non-negative integers and $J = \{j_1, \ldots, j_n\}$ is a multi-index of degree

$$|J| = \sum_{i=1}^{n} j_i = m .$$

In this way we obtain immediately that

$$x^s \frac{\partial h^i}{\partial x^s} = j_s c^i_{j_1 \ldots j_n} (x^1)^{j_1} \ldots (x^n)^{j_n} = j_s c^i_J x^J .$$

On the other hand,

$$A^i_s h^s = \lambda_i c^i_{j_1 \dots j_n} (x^1)^{j_1} \dots (x^n)^{j_n} = \lambda_i c^i_J x^J .$$

Inserting these formulas in (6.20) we obtain

$$\mathcal{L}(h^i) := \lambda_s j_s c^i_J x^J - \lambda_i c^i_J x^J = (\lambda_s j_s - \lambda_i) c^i_J x^J . \qquad (6.22)$$

This shows at once several relevant facts:

1. Each of the vectors v^i_J with all zero components except for the i-th component, given by x^J, is an eigenvector for $\mathcal{L}$, with eigenvalue

$$\sigma(i, J) = (\lambda_s j_s - \lambda_i) .$$

2. The kernel of $\mathcal{L}$ is spanned by the vectors satisfying the *resonance relation*

$$\lambda_i = \lambda_s j_s .$$

3. The range of $\mathcal{L}$ is the space complementary to this kernel; actually one can define a suitable metric (the Bargmann metric, see [19, 55]; see also [11]), such that the range and the kernel are orthogonal.

4. Any function h is written as $c^i_J x^J$; hence the choice of a vector function h_m amounts to the choice of the coefficients c^i_J. In particular, the solution of the homological equation (6.17) is now provided by simple algebraic operations. If we write $f^i_m = \varphi^i_K x^K$ (with K a multi-index of degree m), the Eq. (6.17) reads

$$(\lambda_s j_s - \lambda_i) c^i_J x^J = \pi_\rho [\varphi^i_J x^J] \qquad (6.23)$$

and its solution is obtained by setting all the coefficients c^i_J which correspond to non-resonant terms to

$$c^i_J = \frac{\varphi^i_J}{\lambda_s j_s - \lambda_i} . \qquad (6.24)$$

5. Note that for $\lambda_s j_s \approx \lambda_i$ this will produce *small denominators* and hence very large coefficients c^i_J. If these are larger than $1/\varepsilon$, where ε is the radius of the region (centered in the origin) we are working in, they will make that the series is not well ordered, and cause a number of technical as well as substantial problems. The reader is referred e.g. to [11] for details.

Thus our conclusions in the previous subsection can be restated by saying that (assuming the linear part is associated to a semisimple matrix) *by an ordered sequence of Poincaré transformations one can always reduce the system to one in which all the nonlinear terms are resonant, up to any desired finite order N.*

Note that resonant terms will produce no effect (at least to first order) when acted upon with $\mathcal{L}$; thus the choice of generating functions h_m will always be up to a resonant term.

6.1.2.2 Action of Poincaré Transformations on a Vector Field

We have been discussing the effect of Poincaré transformations on dynamical systems; our discussion can be easily rephrased in terms of vector fields.

In fact, our discussion shows immediately that the Poincaré transformation with generator h_m maps $X = f^i \partial_i$ into the vector field $\widetilde{X} = \widetilde{f}^i \partial_i$. The expression of $\widetilde{f}$ in terms of f and h_m is provided by (6.14) and (6.15) or (6.20).

It is appropriate to note that (6.15) can be immediately interpreted in terms of vector fields. In fact, writing

$$X_k = f_k^i \, \partial_i \tag{6.25}$$

and introducing the vector field

$$Y_m = h_m^i \, \partial_i \, , \tag{6.26}$$

Eq. (6.15) reads

$$\mathcal{L}(h_m) := X_0(h_m) - Y_m(f_0) \, . \tag{6.27}$$

That is, the components of $\mathcal{L}(h_m)$ are the components of the vector field

$$[X_0, Y_m] \, . \tag{6.28}$$

By defining the Lie-Poisson bracket (between vector functions)

$$f, g = f^j \partial_j g^i - g^j \partial_j f^i = (f \cdot \nabla) g - (g \cdot \nabla) f \, , \tag{6.29}$$

we thus just have

$$\mathcal{L}(h_m) = \{f_0, h_m\} \, . \tag{6.30}$$

Moreover, our discussion can now be interpreted by saying that the vector field X is mapped into the vector field $\widetilde{X}$ with $\widetilde{X}_k = X_k$ for $k < m$, and

$$\widetilde{X}_m = X_m - [Y_m, X_0] \, . \tag{6.31}$$

As in the discussion in terms of dynamical systems, the result on higher order components X_k with $k > m$ could be computed but is not of interest here.

We can also restate the result obtained by working in eigen-coordinates (but note this is now coordinate-independent, as we are working with coordinate-independent objects such as vector fields) by saying that *resonant terms are those associated to vector fields which commute with X_0.*

Thus a vector field in normal form is characterized by the fact that *all the nonlinear terms do commute with its linear part.*

Remark 6.2 As commutation is a property which does not depend on changes of coordinates (and the linear part is unaffected by the normalization procedure) we have immediately that, as observed by Moser, normalization can be convergent (and not just a formal procedure) only if the original vector field enjoyed a symmetry. It is maybe surprising that, conversely, the presence of a symmetry can also guarantee the convergence (in a suitably small neighborhood) of the normalization procedure. The reader is referred e.g. to [46, 47], and references therein, for a discussion of these matters.   ⊙

6.1.3 Lie Transforms

We have considered Poincaré transformations, $y = \Phi(x) \equiv x + h(x)$, where $\Phi : M \to M$ is a near-identity diffeomorphism of M; it turns out that it is in several ways convenient to express this as the time-one map for the flow under some vector field H (that is, it is more convenient to deal with such vector fields rather than with finite transformations). We are now going to briefly discuss this approach; the reader is referred e.g. to [20, 24, 46] for more detail.

Let the vector field H be given, in the x coordinates, by $H = h^i(x)\partial_i$. We denote by Ψ the local flow under H, i.e.

$$\frac{\mathrm{d}}{\mathrm{d}s}\Psi(s; x) = H(\Psi(s; x)) ; \quad \Psi(s; x_0) = e^{sH}x_0 . \tag{6.32}$$

We moreover denote $\Psi(1; y)$ as x, i.e. our change of coordinates will be defined to be

$$x = \Psi(1; y) = \left[e^{sH}y\right]_{s=1} ; \tag{6.33}$$

the inverse change of coordinates is simply the reverse flow of the vector field,

$$y = \Psi(-1; x) = \left[e^{-sH}x\right]_{s=1} . \tag{6.34}$$

Now, let us consider the dynamical vector field X. This also generates a (local) flow which we denote by Γ; that is, for any $y_0 \in M$ we have a one-parameter family $y(t) \equiv \Gamma(t; y_0) \in M$ such that $dy/dt = X(y)$.

Combining this with (6.33), we get a one-parameter family $x(t) \in M$ satisfying $dx/dt = \tilde{X}(x)$ for some vector field $\tilde{X} = \tilde{f}^i(x)\partial_i$. We have to determine $\tilde{X}$, i.e. the $\tilde{f}^i$.

We know that $y(t + \delta t) = e^{\delta t X}y(t)$; on the other hand, $y(t) = e^{-H}x(t)$. Hence we have

$$\frac{x(t + \delta t) - x(t)}{\delta t} = \left[e^{sH}\left(e^{(\delta t)X} - I\right)e^{-sH}x(t)\right]_{(s=1)} ,$$

and therefore in conclusion

$$\tilde{X} = \left[e^{sH} X e^{-sH} \right]_{(s=1)} . \tag{6.35}$$

We call this transformation the *Poincaré-Lie transformation* generated by h; or also the *Lie transform* under H.[3]

Note that this yields, up to order one in s—and therefore, if $h = h_k$, up to terms of degree $(k+1)$—just the same transformation as the Poincaré transformation with the same generator h.

The $\tilde{X}$ defined by (6.35) can be given explicitly (for arbitrary s, and up to any order) in terms of the Baker-Campbell-Haussdorf formula [76, 77, 84, 110, 117] (see also [38, 137, 138, 143, 144] for applications of this to differential equations), as

$$\tilde{X} = \sum_{k=0}^{\infty} \frac{(-1)^k s^k}{k!} X^{(k)} , \tag{6.36}$$

where $X^{(0)} = X$ and $X^{(k+1)} = [X^{(k)}, H]$.

As this, or equivalently (6.35), define a transform for any s, we can consider the one-parameter family of vector fields $X_{(s)}$ given by

$$X_{(s)} = f_{(s)}^i(x_{(s)}) \frac{\partial}{\partial x_{(s)}^i} = e^{sH} X e^{-sH} ,$$

where obviously $X_{(0)} = X$, $X_{(1)} = \tilde{X}$. The $f_{(s)}$ will change with s according to

$$df_{(s)}^i/ds = \{h, f_{(s)}\}^i .$$

If $h = h_k$, and we expand f in homogeneous terms (writing $f(s) \equiv f_{(s)}$ to avoid notational confusion), the above formula yields immediately

$$\frac{d f_m(s)}{ds} = \begin{cases} \{h_k, f_{m-k}\} & \text{for } m \geq k \\ 0 & \text{for } m < k . \end{cases}$$

For $m = k$, we just have the action of the homological operator.

Similar considerations would also apply for what concerns Birkhoff normal forms, and Birkhoff transformations, to be considered below.

---

[3] Albeit one should more precisely say it is the action by conjugation of the one parameter group generated by H.

6.1.4 Hamiltonian Dynamical Systems

In the case of Hamiltonian dynamical systems one could of course proceed as for general dynamical systems. However, as observed by Birkhoff, one can more economically deal with the Hamiltonian (i.e. a scalar object) rather than with the associated dynamical system or vector field (which is a vector object).

In fact, recall first that a Hamiltonian is a smooth function $H : M \to \mathbf{R}$, and that to H is associated the vector field X_H on the symplectic manifold (M, ω) defined by

$$X_H \lrcorner \omega = \mathrm{d}H . \tag{6.37}$$

In coordinates, denoting by J the symplectic matrix,[4] we have

$$X = f^i \partial_i ; \quad f^i = J^{ik} \partial_k H . \tag{6.38}$$

As we have seen above, normalization of dynamical systems (or equivalently of vector fields) is based on the Lie-Poisson bracket $\{.,.\}$. It happens that in the algebra of Hamiltonian vector fields (w.r.t. a given symplectic structure) this has a counterpart at the levels of the Hamiltonians, which is of course the standard *Poisson bracket*.

In fact, let $X = f^i \partial_i$ and $Y = g^i \partial_i$ be Hamiltonian vector fields with Hamiltonian respectively F and G. Then we have, using the fact J is constant and skew-symmetric, and with an integration by parts,

$$
\begin{aligned}
h^i = \{f, g\}^i &= f^j \, (\partial_j g^i) - g^j \, (\partial_j f^i) \\
&= (J^{jk} \, \partial_k F) \, [\partial_j (J^{im} \partial_m G)] - (J^{jk} \, \partial_k G) \, [\partial_j (J^{im} \partial_m F)] \\
&= J^{jk} \, (\partial_k F) \, J^{im} (\partial^2_{jm} G) - J^{jk} \, (\partial_k G) \, J^{im} \, (\partial^2_{jm} F) \\
&= J^{im} \, \partial_m [(\partial_k F) \, J^{jk} \, (\partial_j G)] \\
&\quad - J^{im} (\partial^2_{mk} F) \, J^{jk} \, (\partial_j G) - J^{jk} \, (\partial_k G) \, J^{im} \, (\partial^2_{jm} F) \\
&= J^{im} \, \partial_m [(\partial_k F) \, J^{jk} \, (\partial_j G)] \\
&\quad + J^{im} (\partial^2_{mk} F) \, J^{kj} \, (\partial_j G) - J^{im} \, (\partial^2_{mj} F) \, J^{jk} \, (\partial_k G) \\
&= J^{im} \, \partial_m [(\partial_k F) \, J^{jk} \, (\partial_j G)] = J^{im} \, (\partial_m H) .
\end{aligned}
$$

In the last step we have of course defined

$$H = [(\partial_k F) \, J^{jk} \, (\partial_j G)] := \{F, G\} , \tag{6.39}$$

where $\{F, G\}$ is the standard Poisson bracket of the functions F and G.

---

[4]By Darboux theorem [10, 12, 30, 82] this can always be reduced locally to the standard form; in particular, it can be brought to be *constant* in convenient (Darboux) coordinates.

Thus we have shown that the Lie-Poisson bracket of the Hamiltonian vector fields associated to the Hamiltonians F and G is the Hamiltonian vector field associated to the Hamiltonian which is the Poisson bracket of F and G, $H = \{F, G\}$,

$$\{X_F, X_G\} = X_{\{F,G\}} .$$

This result allows at once to deal with Hamiltonians rather than with associated vector fields; in particular, we will be able to simplify the expression for the dynamics working directly at the level of the Hamiltonians.

6.1.5 Birkhoff-Gustavsson Normal Forms

In the case of Hamiltonian systems, we have a manifold M of dimension $m = 2n$ equipped with a symplectic form. We will have a Hamiltonian H with a non-degenerate fixed point—i.e. a minimum, as we want a stable fixed point—in the point m_0. This will be chosen as the origin in the system of (Darboux) local coordinates $(x^1, \ldots, x^{2n}) = (p_1, \ldots, p_n; q^1, \ldots, q^n)$, and in these the symplectic form is described by the matrix

$$J = \begin{pmatrix} 0 & -I_n \\ I_n & 0 \end{pmatrix} .$$

The hypothesis that $x = 0$ is a non-degenerate minimum of H means that (by adding an inessential constant to H) we can always set

$$H(0) = 0 ; \quad (\nabla H)(0) = 0 . \tag{6.40}$$

Thus we expand H in a Taylor series around $x = 0$; this will be written as

$$H(x) = \sum_{k=0}^{\infty} H_k(x) , \tag{6.41}$$

where H_k is homogeneous of degree $(k + 2)$.

As for the Poincaré transformations, these will be canonical maps with generators F_m (homogeneous of degree $m + 2$). They will leave unchanged the terms H_k with $k < m$, while the term H_m will change[5] according to

$$H_m \rightarrow \widetilde{H}_m = H_m + \{F_m, H_0\} ; \tag{6.42}$$

this is the Hamiltonian expression of (6.14).

---

[5] As in the general case, here H_m should be meant as the expression taken by H_m after the previous normalization steps have been performed. That is, at each step after computing the $\widetilde{H}_k$ we will rename them as H_k.

Thus we have an expression for the homological operator acting on Hamiltonians; that is,

$$\mathcal{L} = \{H_0, .\} . \tag{6.43}$$

Note this will map Hamiltonians which are homogeneous of degree $(m + 2)$ into Hamiltonians homogeneous of the same degree. Thus we can consider the restrictions $\mathcal{L}_k$ to the space of homogeneous functions of degree $(k+2)$; this is a finite dimensional space and hence $\mathcal{L}_k$ can be represented as a finite dimensional matrix.

Note also that H_0 is identified by a (real) matrix A, which can always be assumed to be symmetric:

$$H_0(x) = \frac{1}{2} A_{ij} x^i x^j \qquad A_{ji} = A_{ij} .$$

The eigenvalues of this will always come in pairs $\pm i\omega$, which makes that there will always be trivial resonances.

Resonant terms F will be identified by

$$\{H_0, F\} = 0 . \tag{6.44}$$

The normalization procedure—which in this context takes the name of Birkhoff-Gustavsson normalization[6]—will allow to eliminate all terms which are in the range of $\mathcal{L}$, i.e. all non-resonant terms. We will thus arrive to the Birkhoff-Gustavsson normal form

$$\widetilde{H} = H_0 + \sum_{k=1}^{\infty} \widetilde{H}_k , \tag{6.45}$$

where all the $\widetilde{H}_k$ are resonant.

6.2   Perturbation of Hyperhamiltonian Systems

We will now try to adopt the same point of view for hyperhamiltonian systems. An important point to stress is that we should and will *not* consider perturbations of a generic hyperhamiltonian system, but instead (as in the generic and in the Hamiltonian cases) perturbations of an *integrable* hyperhamiltonian system (moreover, as inherent to the normal forms approach, a linear one). As we have seen in Chap. 5, this means perturbation of a system of *quaternionic oscillators*.

We will again work in the neighborhood $U \subseteq M$ of some fixed point m_0 in the $4n$-dimensional manifold M, and choose local coordinates x^i $(i = 1, \ldots, 4n)$ in U.

---

[6]Quite similarly to the Poincaré-Dulac case, here Birkhoff worked out the case of nonresonant (Hamiltonian) systems, and Gustavsson extended the result to the resonant case. Note that here by "non resonant" it is meant there are no resonances beside the trivial ones (corresponding to eigenvalues coming in pairs of complex conjugate ones, see above).

Most of our discussion will be conducted in the Euclidean framework; thus the hyperkahler structure is a standard one (see Sect. 1.4.3) and it corresponds to *constant* (rather than covariantly constant) matrices.

6.2.1 Expansion of Hyperhamiltonian Vector Fields

Our vector field and the associated dynamical system will be written as in (6.1), (6.2), i.e. as

$$X = f^i(x) \, \partial_i \, ,$$

or equivalently as the ODE

$$\dot{x}^i = f^i(x) \, ; \tag{6.46}$$

but now we assume to have a hyperhamiltonian system. Thus there are Hamiltonians $\mathcal{F}^\alpha$ such that

$$f^i(x) = \sum_\alpha M_\alpha^{ij} \, (\partial_j \mathcal{F}^\alpha) \, . \tag{6.47}$$

As for the hyperkahler structure, as anticipated, we will deal with the case where (at least in U, if not in all of M) this is the standard one. That is, we have a Euclidean metric $g = I_{4n}$ and a standard hyperkahler structure, with the matrices Y_α, K_α and M_α seen in Sect. 1.4.3. If we deal with $(M, g) = (\mathbf{R}^{4n}, \delta)$ we can of course take $U = M$.

Dealing with the standard hyperkahler structures means in particular that in (6.47) we can think the M_α as *constant* matrices (see the detailed discussion given in [68, 70]).

Thus when expanding f in a Taylor series, our expansion actually only deals with the Hamiltonians. We will write, as in the standard Hamiltonian case,

$$\mathcal{F}^\alpha(x) = \sum_{k=0}^{\infty} \mathcal{F}_k^\alpha(x) \, , \tag{6.48}$$

with $\mathcal{F}_k^\alpha$ homogeneous of degree $(k + 2)$ in the x^i. Thus when expanding f we will obtain

$$f^i(x) = \sum_{k=0}^{\infty} f_k^i(x) = \sum_{k=0}^{\infty} \sum_{\alpha=1}^{3} M_\alpha^{ij} \, (\partial_j \mathcal{F}_k^\alpha) \, . \tag{6.49}$$

Note that while the scalar function $\mathcal{F}_k^\alpha$ is homogeneous of degree $(k + 2)$, the corresponding vector function f_k is homogeneous of degree $(k + 1)$.

Remark 6.3 The linear term f_0, corresponding to the quadratic Hamiltonians $\mathcal{F}_0^\alpha$, is special. In fact, as we have seen in Sect. 5.1, for a set of quadratic Hamiltonians we can always (by an SO(3) rotation in the Kahler sphere) reduce to the standard Hamiltonian case. Thus the linear term is always—or can always taken to be—in standard Hamiltonian form. In other words, we can always take $\mathcal{F}_0^2 = \mathcal{F}_0^3 = 0$, and correspondingly $f_0^i = M_1^{ij} \partial_j \mathcal{F}_0^1$. The nonlinear terms f_k ($k > 1$) will instead be, in general, of truly hyperhamiltonian nature.   ⊙

6.2.2 Lie Transforms for Hyperhamiltonian Vector Fields

We will now consider a change of coordinates; we will of course still consider one generated by a Lie transform under the flow of the vector field

$$Y = g^i(x) \, \partial_i \, . \tag{6.50}$$

Similarly to what is done in the Hamiltonian case, here Y will *not* be a general vector field, but instead one of the "same nature" as X. This should guarantee that the transformed vector field $\widetilde{X}$ will have the same properties of the original one, in particular it will be canonical for the hyperkähler structure.

Remark 6.4 The notion of "same nature" deserves a pause for discussion. The first natural option is of course to take for Y a hyperhamiltonian vector field as well, thus

$$g^i = M_\alpha^{ij} \, (\partial_i \mathcal{G}^\alpha) \tag{6.51}$$

for some triple of Hamiltonians $\mathcal{G}^\alpha$.

Under reflection, however, another option appears equally natural: that is, considering Y a hyperhamiltonian vector field for a *dual* hyperkähler structure, see Sect. 1.5.

In fact, as we have seen in Sect. 3.2, both vector fields which are hyperhamiltonian w.r.t. the given hyperkähler structure and those which are hyperhamiltonian w.r.t. the dual ones generate canonical transformations. Actually, these are all the canonical maps generators, see Sect. 3.3 and [71].

In this case we would have

$$g^i = \widehat{M}_\alpha^{ij} \, (\partial_i \widehat{\mathcal{G}}^\alpha) \tag{6.52}$$

for some triple of Hamiltonians $\widehat{\mathcal{G}}^\alpha$ and matrices $\widehat{M}_\alpha$ associated to a given dual structure.

A third, more general option, is of course also possible: that is, considering a superposition of hyperhamiltonian vector fields w.r.t. the original hyperkähler structure and all of its dual ones (that is, a Dirac vector field, see Sect. 3.3). We will write in this case

$$g^i = M_\alpha^{ij} \, (\partial_i \mathcal{G}^\alpha) + \widehat{M}_\beta^{ij} \, (\partial_i \widehat{\mathcal{G}}^\beta) \, , \tag{6.53}$$

where the notation should be obvious. Note indeed that for any hyperkähler structure dual to the given one, denote by $\widetilde{M}_\alpha$ the associated matrices, on each irreducible hyperkähler submanifold these are either coinciding with the M_α or with the $\widehat{M}_\alpha$, so upon rearrangement we are always reduced to this situation.

For further details on dual hyperkahler structures, Dirac vector fields and their properties in connection with canonical and **Q**-maps, the reader is referred to Sects. 1.5, 2.4.2 and 3.3.   ⊙

Remark 6.5 If we want to be able to consider these different options at the same time in our subsequent computations, we will write

$$g^i \; = \; Q^{ij}_\beta \; (\partial_i \mathcal{G}^\beta) \; ; \tag{6.54}$$

here we can have $\beta = 1, 2, 3$ and $Q_\beta = M_\beta$ or $Q_\beta = \widehat{M}_\beta$; or even $\beta = 1, \ldots, 6$ with $Q_\beta = M_\beta$ for $\beta = 1, 2, 3$ and $Q_\beta = \widehat{M}_{\beta-3}$ for $\beta = 4, 5, 6$. In the following we will make use of this notation, which will allow to work in the "natural" setting mentioned at the beginning of Remark 6.4 (that is, just hyperhamiltonian vector fields) and later on discuss the different results which are obtained with the different choices for Y without having to repeat the various computations.[7]

Note that this notation also allows to consider at the same time several dual structures, i.e. a general Dirac vector field, without the need for the rearrangement mentioned at the end of Remark 6.4, just by extending the range of the index β, and considering matrices Q_β associated to all of these dual structures.   ⊙

6.2.2.1   The Coordinate Approach. Explicit Formulas

We can now come back to consider the action of the Lie transform generated by the vector field Y on X. Working at the level of vector fields, we can use the general formulas seen in Sect. 6.1.1. We thus have to evaluate the commutator $Z = [X, Y]$ or, working in coordinates, the Lie-Poisson bracket

$$h \; := \; \{f, g\} \; ;$$

obviously we would have $Z = h^i \partial_i$.

Ideally, we should show that this is still a hyperhamiltonian vector fields (w.r.t. the same hyperkahler structure), i.e. there are Hamiltonians $\mathcal{H}^\alpha$ such that

$$h^i \; = \; M^{ij}_\alpha \, \partial_j \mathcal{H}^\alpha \, , \tag{6.55}$$

and possibly obtain an explicit expression for the $\mathcal{H}^\alpha$ in terms of the $\mathcal{F}^\alpha$ and $\mathcal{G}^\alpha$. Unfortunately this is not the case in such a generality.

This is a direct computation which will make use of an integration by parts and of the fact M_α and Q_α are constant. We will also use a short notation for second derivatives, i.e. write $F^\alpha_{ij} := \partial^2_{ij} \mathcal{F}^\alpha$ (and similarly $G^\alpha_{ij} := \partial^2_{ij} \mathcal{G}^\alpha$). We obtain[8]

---

[7]Note also that with such a notation we can discuss at the same time also the problem of Lie transforms acting on a Dirac vector field, simply extending the range of the summation index α relative to the f; see below for this extension.

[8]The reason for the seemingly odd notation $\widetilde{\mathcal{H}}^\beta_{(0)}$ (note here the subscript "(0)" does not refer to homogeneity degree) will be clear in the following.

$$h^i = \{f, g\}^i = f^j (\partial_j g^i) - g^j (\partial_j f^i)$$
$$= M_\alpha^{j\ell} (\partial_\ell \mathcal{F}^\alpha) \, \partial_j [Q_\beta^{im} (\partial_m \mathcal{G}^\beta)] - Q_\beta^{j\ell} (\partial_\ell \mathcal{G}^\beta) \, \partial_j [M_\alpha^{im} (\partial_m \mathcal{F}^\alpha)]$$
$$= Q_\beta^{im} \partial_m [(\partial_j \mathcal{G}^\beta) M_\alpha^{j\ell} (\partial_\ell \mathcal{F}^\alpha)]$$
$$- M_\alpha^{j\ell} F_{\ell m}^a Q_\beta^{im} (\partial_j \mathcal{G}^\beta) - Q_\beta^{j\ell} (\partial_\ell \mathcal{G}^\beta) M_\alpha^{im} F_{jm}^\alpha$$
$$= Q_\beta^{im} \partial_m \tilde{\mathcal{H}}_{(0)}^\beta + \Delta^i \, .$$

In the final writing we have separated the term which is obviously in hyperhamiltonian form, and defined the Hamiltonians

$$\tilde{\mathcal{H}}_{(0)}^\beta := (\partial_j \mathcal{G}^\beta) \, M_\alpha^{j\ell} \, (\partial_\ell \mathcal{F}^\alpha) \, , \tag{6.56}$$

on the one hand; and the term Δ^i which still needs discussion on the other hand. This is rewritten (recalling that the M_α and Q_β are skew-symmetric matrices, while F^α and G^β are symmetric) as

$$\Delta^i = Q_\beta^{im} F_{m\ell}^\alpha M_\alpha^{\ell j} \partial_j \mathcal{G}^\beta - M_\alpha^{im} F_{mj}^\alpha Q_\beta^{j\ell} \partial_\ell \mathcal{G}^\beta$$
$$= [Q_\beta^{im} F_{m\ell}^\alpha M_\alpha^{\ell j} - M_\alpha^{im} F_{m\ell}^\alpha Q_\beta^{\ell j}](\partial_j \mathcal{G}^\beta)$$
$$= \left[(Q_\beta F^\alpha M_\alpha - M_\alpha F^\alpha Q_\beta) \nabla G^\beta \right]^i$$
$$:= [P_\beta \nabla G^\beta]^i \, ,$$

where we have defined the (skew-symmetric) matrices[9]

$$P_\beta := (Q_\beta F^\alpha M_\alpha - M_\alpha F^\alpha Q_\beta) \, , \tag{6.57}$$

and we should recall that the $\mathcal{G}^\alpha$ (and hence their gradients) are arbitrary. This latter remark means that proving that Δ^i always vanish would require to show that the P_β are always zero; this is clearly not true.

Recall however we want to prove that $Z = [X, Y] = h^i \partial_i$ is hyperhamiltonian. This will be true not only if $\Delta^i = 0$, but also if the Δ^i can be written in hyperhamiltonian form, i.e. if there are $\tilde{\mathcal{H}}_{(1)}^\alpha$ such that

$$\Delta^i = M_\alpha^{ij} \partial_j \tilde{\mathcal{H}}_{(1)}^\alpha \, . \tag{6.58}$$

A look at the explicit expression for the Δ^i shows that it is not easy to determine if this is the case; one can actually prove it cannot be true in general.

In fact, consider the case where $\mathcal{G}^2 = \mathcal{G}^3 = 0$; then focusing e.g. on P_2, i.e. requiring $P_2 = c_\alpha M_\alpha$, we get a solution—with $c_1 = c_2 = 0$ and $c_3 = -(k_1 + k_3)$—if and only if G^1 is of the form

---

[9]In general non constant; see also Remark 6.6 below.

$$G^1 = \begin{pmatrix} K & k_4 & k_5 & -k_6 \\ k_4 & k_1 & k_6 & k_7 \\ k_5 & k_6 & k_2 & k_4 \\ -k_6 & k_7 & k_4 & k_3 \end{pmatrix} \; ; \quad K := k_1 - k_2 + k_3 \; .$$

As this is *not* the general form of G^1 (which is only required to be symmetric), we conclude that in general the situation is not the one we would like to have.

However, *a special situation occurs when the $\mathcal{F}^\alpha$ correspond to an integrable hyperhamiltonian system*, i.e. to a *quaternionic oscillator*, as discussed in the next Section.

Remark 6.6 For the term Δ computed above to be in Dirac form we would need that the matrix P_β defined in (6.57) is a linear combination of the hyperkahler matrices for the given and the dual hyperkahler structures. This is in general not guaranteed, as can be seen already in the Euclidean four-dimensional case. In fact, the matrix P is skew-symmetric, and hence can surely be expressed in terms of the M and the $\widehat{M}$ matrices; but these are constant (and in the general case are covariantly constant), while P is in general—due to the presence of the F^α terms—not constant (nor covariantly constant). That is, we can always write

$$P_b = \sum_\alpha \left(W_{\beta\alpha} M_\alpha + \widehat{W}_{\beta\alpha} \widehat{M}_\alpha \right) \; ,$$

but the W, $\widehat{W}$ are general smooth functions and not constant coefficients (or, in the general case, covariantly constant functions). $\odot$

Remark 6.7 The expression (6.57) for P_β is not symmetric, in that only the F^α, and not the G^α as well, appear in it. Similarly, in (6.56) the role of the F^α and of the G^α is not equivalent. One can check that it is also possible to obtain a symmetric expression for P_α; this also corresponds to a different, and more symmetric, expression for the $\widetilde{\mathcal{H}}^\alpha_{(0)}$. However, in the following Sect. 6.3 it will be essential to have an asymmetric role for the F^α and the G^α. $\odot$

Remark 6.8 Note that (6.56) can be interpreted in terms of the Poisson brackets $\{.,.\}_\alpha$ associated to the three symplectic structures ω_α. In fact, in these terms we have

$$\widetilde{\mathcal{H}}^\alpha_{(0)} = \left(\frac{\partial \mathcal{F}^\beta}{\partial x^\ell} M^{\ell k}_\beta \frac{\partial \mathcal{G}^\alpha}{\partial x^k} \right) = \sum_{\beta=1}^3 \{\mathcal{F}^\beta, \mathcal{G}^\alpha\}_\beta \; . \tag{6.59}$$

Here in the r.h.s. we have written explicitly the sum on β (involving the index of the Hamiltonian and that for the Poisson bracket, i.e. the symplectic structure) for greater clarity. $\odot$

6.3 Lie Transforms and Quaternionic Oscillators

The situation considered in the previous Section becomes quite simpler when one of the two sets of Hamiltonians is quadratic. Note that we have already considered a formulation in which the roles of the F^α and the G^α Hamiltonians is *not* symmetric, see Remark 6.7. We are thus ready to take advantage of the special properties of quadratic Hamiltonians.

As usual, it will be convenient to deal with the simplest ($n = 1$) case at first, and then consider the case of general $4n$-dimensional spaces.

6.3.1 Simple Linear Quaternionic Oscillators ($n = 1$)

Let us consider for a moment the case of four-dimensional manifolds, for ease of notation. If the $\mathcal{F}$ define a quaternionic oscillator, it means the $\mathcal{F}^\alpha(x)$ depend on x only through $\rho = (1/2)|\mathbf{x}|^2$, and therefore

$$F_{ij}^\alpha \;=\; \frac{\partial^2 \mathcal{F}^\alpha}{\partial x^i \partial x^j} \;=\; x^i\, x^j\, \frac{d^2 \mathcal{F}^\alpha}{d\rho^2} \;+\; \delta_{ij}\, \frac{d\mathcal{F}^\alpha}{d\rho} \,. \tag{6.60}$$

If moreover the $\mathcal{F}^\alpha$ are quadratic in the x, hence *linear* in ρ, which corresponds to a linear quaternionic oscillator

$$\mathcal{F}^\alpha \;=\; \mu^\alpha\, \rho$$

(here the μ^α are real constants), the first term in the r.h.s. of (6.60) vanishes, and we are left with

$$F_{ij}^\alpha \;=\; \delta_{ij}\, \frac{d\mathcal{F}^\alpha}{d\rho} \;:=\; \delta_{ij}\, \mu^\alpha. \tag{6.61}$$

In this case the P_β are therefore

$$P_\beta \;:=\; \mu^\alpha\, (Q_\beta M_\alpha - M_\alpha Q_\beta)\,. \tag{6.62}$$

It suffices now to recall that $[M_\alpha, \widehat{M}_\beta] = 0$ and

$$[M_\alpha, M_\beta] \;=\; 2\, \epsilon_{\alpha\beta\gamma}\, M_\gamma$$

to conclude that (under our assumptions) any vector field which is hyperhamiltonian w.r.t. the dual hyperkahler structure will give a zero contribution to the P_β and hence to Δ^i, while vector fields which are hyperhamiltonian w.r.t. the same hyperkahler structure as X (hence with $Q_\beta = M_\beta$) will yield

$$P_\beta \;=\; 2\, \epsilon_{\beta\alpha\gamma}\, \mu^\alpha\, M_\gamma$$

and hence

$$\Delta^i = P_\beta^{ij} \, (\partial_j \mathcal{G}^\beta)$$
$$= 2 \, \epsilon_{\beta\alpha\gamma} \, \mu^\alpha M_\gamma^{ij} (\partial_j \mathcal{G}^\beta)$$
$$= M_\gamma^{ij} \, \partial_j \left(2 \, \epsilon_{\beta\alpha\gamma} \, \mu^\alpha \, \mathcal{G}^\beta \right) \ .$$

Direct comparison with (6.58) and a rearrangement of indices show that

$$\widetilde{\mathcal{H}}_{(1)}^\alpha \; = \; - \, 2 \, \epsilon_{\alpha\beta\gamma} \, \mu^\beta \, \mathcal{G}^\gamma \ . \tag{6.63}$$

This completes the proof that (in dimension four, and with our hypotheses about $\mathcal{F}^\alpha$) Eq. (6.55) holds; the Hamiltonians $\mathcal{H}^\alpha$ are given by

$$\mathcal{H}^\alpha \; = \; \widetilde{\mathcal{H}}_{(0)}^\alpha + \widetilde{\mathcal{H}}_{(1)}^\alpha \; = \; \mu^\alpha \, x^j \, Q_\beta^{jm} \, (\partial_m \mathcal{G}^\beta) + 2 \, \epsilon_{\alpha\beta\gamma} \, \mu^\gamma \, \mathcal{G}^\beta \ . \tag{6.64}$$

Remark 6.9 Note that if the $\mathcal{G}^\alpha$ are homogeneous of degree k, then the $\mathcal{H}^\alpha$ are also homogeneous of the same degree k. $\qquad\qquad\qquad\qquad\qquad\qquad\qquad\qquad\odot$

Remark 6.10 We stress again that, as mentioned in Sect. 5.1 and Remark 6.3, linear quaternionic oscillators can always be brought to Hamiltonian form. On the other hand, here we are considering *general* hyperhamiltonian (and actually, Dirac) perturbations of such a system, thus we go beyond the realm of Hamiltonian perturbations. $\qquad\qquad\qquad\qquad\qquad\qquad\qquad\qquad\qquad\qquad\qquad\qquad\odot$

6.3.2 General Linear Quaternionic Oscillators (n > 1)

We have so far performed our computations assuming to be in the four-dimensional Euclidean space $\mathbf{R}^4$. Extension of these to Euclidean $\mathbf{R}^{4n}$ is rather straightforward; the discussion is greatly eased by use of the block notation introduced in Sect. 5.1, which we will in fact use below.

That is, we will consider n four-dimensional vectors $\xi_{(k)}$, ($k = 1, \ldots, n$) with components $\xi_{(k)}^i = x^{4(k-1)+i}$. We will also use variables $\rho_{(k)} := (1/2)|\xi_{(k)}|^2$. In some places we will denote by k_i the block to which x^i belongs; that is

$$k_i \; := \; [(i-1)/4 + 1] \tag{6.65}$$

(here the square brackets denote integer part).

Our general formulas remain true; in particular we still have that

$$h^i = Q^{im}_\beta \, \partial_m \tilde{\mathcal{H}}^\beta_{(0)} + \Delta^i \; ;$$
$$\tilde{\mathcal{H}}^\beta_{(0)} = (\partial_j \mathcal{F}^\alpha) \, Q^{jm}_\beta \, (\partial_m \mathcal{G}^\beta) \; ,$$
$$\Delta^i = P^{ij}_\beta \, \partial_j \mathcal{G}^\beta \; ,$$
$$P_\beta = \left(Q_\beta \, \mathcal{F}^\alpha \, M_\alpha \; - \; M_\alpha \, \mathcal{F}^\alpha \, Q_\beta \right) \; .$$

Now the assumption that X correspond to an integrable system, i.e. a (multidimensional) quaternionic oscillator, results in

$$\mathcal{F}^\alpha(x) \; = \; \mathcal{F}^\alpha \left(\rho_{(1)}, \ldots, \rho_{(k)} \right) \; . \tag{6.66}$$

For such $\mathcal{F}^\alpha$, the first derivatives read

$$\frac{\partial \mathcal{F}^\alpha}{\partial x^i} \; = \; \frac{\partial \mathcal{F}^\alpha}{\partial \rho_{(k_i)}} \, x^i \; ; \tag{6.67}$$

as for second derivatives, formula (6.60) is replaced by

$$F^\alpha_{ij} \; = \; \frac{\partial^2 \mathcal{F}^\alpha}{\partial x^i \partial x^j} \; = \; x^i \, x^j \, \frac{\partial^2 \mathcal{F}^\alpha}{\partial \rho_{(k_i)} \, \partial \rho_{(k_j)}} \; + \; \delta_{ij} \, \frac{d \mathcal{F}^\alpha}{d \rho_{(k_i)}} \; . \tag{6.68}$$

If we moreover require that the $\mathcal{F}^\alpha$ are quadratic in the x, hence *linear* in $\rho_{(k)}$, so that we have a *linear* quaternionic oscillator (see Remark 6.3 in this respect), this results in that the Hamiltonians $\mathcal{F}^\alpha$ must simply be of the form

$$\mathcal{F}^\alpha(x) \; = \; \sum_{i=1}^n \mu^\alpha_{(k)} \, \rho_{(k)} \; . \tag{6.69}$$

For these $\mathcal{F}^\alpha$, we have just

$$\frac{\partial \mathcal{F}^\alpha}{\partial x^i} \; = \; \mu^\alpha_{(k_i)} \, x^i \tag{6.70}$$

for first derivatives. As for second derivatives, we get

$$F^\alpha_{ij} \; = \; \delta_{ij} \, \frac{\partial \mathcal{F}^\alpha}{\partial \rho_{(k_i)}} \; := \; \delta_{ij} \, \mu^\alpha_{(k_i)} \; ; \tag{6.71}$$

this replaces formula (6.61) in the n-dimensional case.

We can now compute $\widetilde{\mathcal{H}}^\alpha_{(0)}$ and Δ^i (and from this $\widetilde{\mathcal{H}}^\alpha_{(1)}$) in the n-dimensional case. In fact, just by substituting the expressions above in our general formula, we easily get

$$\widetilde{\mathcal{H}}^\alpha_{(0)} \;=\; \mu^\alpha_{(k_j)}\, x^j\, Q^{jm}_\beta\, (\partial_m \mathcal{G}^\beta)\;. \tag{6.72}$$

As for P_β, these will be skew-symmetric block-diagonal matrices; we denote by $(P_\beta)_{(k)}$ the k-th (four-dimensional) block on these, and similarly for the M_α and Q_β matrices (which are themselves block-diagonal). With this notation we have

$$(P_\beta)_{(k)} \;=\; \mu^\alpha_{(k)}\, \left(Q_\beta\, M_\alpha \;-\; M_\alpha\, Q_\beta\right)_{(k)}\;, \tag{6.73}$$

so that on each block the same considerations as in the four-dimensional case apply. In particular, for $Q_\beta = \widehat{M}_\beta$ (or more precisely $Q_\beta = \widehat{M}_{\beta-3}$) we have

$$\left([\widehat{M}_\beta, M_\alpha]\right)_{(k)} \;=\; 0\;, \tag{6.74}$$

while for $Q_\beta = M_\beta$ we get

$$(P_\alpha)_{(k)} \;=\; 2\, \epsilon_{\alpha\beta\gamma}\, \mu^\beta_{(k)}\, \left(M_\gamma\right)_{(k)}\;. \tag{6.75}$$

We will write, for ease of notation, Φ^β for the block-diagonal matrix with block entries $\mu^\beta I_4$; note that with this notation Eq. (6.72) reads

$$\widetilde{\mathcal{H}}^\alpha_{(0)} \;=\; (\Phi^\alpha x)^j\, Q^{jm}_\beta\,(\partial_m \mathcal{G}^\beta) \;=\; \left((\Phi^a x)^i\, Q^{ij}_\beta\,(\partial_j \mathcal{G}^\beta)\right)\;. \tag{6.76}$$

Then the above discussion shows that Δ^i is given by

$$\Delta^i_{(k)} \;=\; 2\, \epsilon_{\alpha\beta\gamma}\, \Phi^\beta\, M^{im}_\gamma\,(\partial_m \mathcal{G}^\alpha) \;=\; M^{im}_\gamma\, [\partial_m (2\, \epsilon_{\gamma\alpha\beta}\mu^\beta_{(k)}\, \mathcal{G}^\alpha)]\;; \tag{6.77}$$

it follows from this that

$$\widetilde{\mathcal{H}}^\alpha_{(1)} \;=\; -2\, \epsilon_{\alpha\beta\gamma}\, \Phi^\beta\, \mathcal{G}^\gamma\;. \tag{6.78}$$

In conclusion, we get

$$\mathcal{H}^\alpha \;=\; \widetilde{\mathcal{H}}^\alpha_{(0)} \;+\; \widetilde{\mathcal{H}}^\beta_{(1)} \;=\; \left(\Phi^\alpha x,\, Q_\beta \nabla \mathcal{G}^\beta\right) \;-\; 2\, \epsilon_{\alpha\beta\gamma}\, \Phi^\beta \mathcal{G}^\gamma\;. \tag{6.79}$$

6.3.3 Discussion and Recapitulation of the Results

It is convenient to summarize the main result of our discussion in a couple of formal statements.

Lemma 6.1 *Let M be a hyperkahler manifold; let X be the hyperhamiltonian vector field generated by the Hamiltonians $\mathcal{F}^\alpha$, and suppose these are, with the notation introduced above, of the form*

$$\mathcal{F}^\alpha(x) = \sum_k \mu^\alpha_{(k)}\, \rho_{(k)}\,,$$

with $\mu^\alpha_{(k)}$ real constants (hence X is linear). Let Y be the hyperhamiltonian vector field (w.r.t. the same hyperkahler structure) generated by general smooth Hamiltonians $\mathcal{G}^\alpha$.

Then the vector field $Z := [X, Y]$ is hyperhamiltonian (w.r.t. the same hyperkahler structure) with Hamiltonians $\mathcal{H}^\alpha$ given by

$$\mathcal{H}^\alpha = \left((\Phi^\alpha)_{ij}x^j\right)\, Q^{im}_\beta\, (\partial_m \mathcal{G}^\beta) - 2\,\epsilon_{\alpha\beta\gamma}\Phi^\beta\, \mathcal{G}^\gamma\,. \tag{6.80}$$

As mentioned in Remarks 6.4 and 6.5, our notation allows to consider a more general setting, that is to include at the same time vector fields which are hyperhamiltonian with respect to a pair of dual hyperkahler structures. Thus, with exactly the same computations (at the formal level, but with different interpretation; see Remark 6.5 above) we obtain an extension of Lemma 6.1. In this, the vanishing of one or the other of the different triples of Hamiltonians considered allows to obtain special versions of the same result.

In the following statement, we write $M^\alpha_{(+)} = M_\alpha$, $M^\alpha_{(-)} = \widehat{M}_\alpha$, and $\Phi^\alpha_\pm$ is the matric Φ built with the $\mu^\alpha_\pm$ corresponding to the $\mathcal{F}^\alpha_\pm$ quadratic Hamiltonians.

Lemma 6.2 *Let M be a hyperkahler manifold; let $X_\pm$ be the hyperhamiltonian vector field generated by the Hamiltonians $\mathcal{F}^\alpha_\pm$ with respect to a pair of dual hyperkahler structures, and suppose these are, with the notation introduced above, of the form $\mathcal{F}^\alpha_\pm(x) = \mu^\alpha_{\pm,(k)}\rho_{(k)}$, with $\mu^\alpha_{\pm,(k)}$ real constants. Let $Y_\pm$ be the hyperhamiltonian vector field generated by general smooth Hamiltonians $\mathcal{G}^\alpha_\pm$ w.r.t. the same pair of dual hyperkahler structures. Write $X = X_+ + X_-$, and similarly $Y = Y_+ + Y_-$*

Then the vector field $Z := [X, Y]$ is a superposition of hyperhamiltonian vector fields $Z_\pm$ (w.r.t. the same pair of dual hyperkahler structures) with Hamiltonians $\mathcal{H}^\alpha_\pm$ given by

$$\mathcal{H}^\alpha_\pm = (\Phi^\beta_\pm x)_\ell \left[(M^{(\pm)}_\alpha)^{\ell k}\partial_k \mathcal{G}^\beta_\pm + (M^{(\mp)}_\alpha)^{\ell k}\partial_k \mathcal{G}^\beta_\mp\right] - 2\,\epsilon_{\alpha\beta\gamma}\,(\Phi^\beta_\pm)\,\mathcal{G}^\gamma_\pm\,. \tag{6.81}$$

Remark 6.11 If in (6.80) or (6.81) the $\mathcal{G}^\alpha$ are all homogeneous of the same degree k, then the $\mathcal{H}^\alpha$ are also homogeneous of the same degree k. ⊙

It is interesting to note that (6.79), or the formulas (6.80) and (6.81) appearing in the two Lemmas, can be interpreted in terms of a triple of *linear* operators— associated to Φ^α and hence to the quadratic terms $\mathcal{H}_0^\alpha$ in the Hamiltonians $\mathcal{H}^\alpha$— acting on triples of functions $\mathcal{G}^\alpha$ which we write as

$$\mathcal{L}^\alpha(\mathcal{G}) := [(\Phi^\alpha)^i{}_j x^j]\, Q_\beta^{jk}\, \partial_k \mathcal{G}^\alpha \;-\; 2\, \epsilon_{\alpha\beta\gamma}\, \Phi^\beta\, \mathcal{G}^\gamma \,. \tag{6.82}$$

This is the *homological operator* for hyperhamiltonians and hyperhamiltonian dynamics.

Moreover, Remark 6.11 shows that we can consider the restriction $\mathcal{L}_{(k)}^\alpha$ to the (finite-dimensional) space $\mathbf{H}_{(k)}$ of Hamiltonians which are homogeneous of degree $k+2$, as

$$\mathcal{L}^\alpha \;:\; \mathbf{H}_{(k)} \;\to\; \mathbf{H}_{(k)} \,.$$

Proceeding as in the standard Poincaré-Birkhoff normalization, at each step with generating functions $\mathbf{G}_m = \{\mathcal{G}_m^1, \mathcal{G}_m^2, \mathcal{G}_m^3\}$, where of course $\mathcal{G}_m^\alpha \in \mathbf{H}_{(m)}$, the terms $\mathcal{F}_k^\alpha$ with $k < m$ will stay unchanged, those with $k > m$ will change in a way we do not need to control, and the term $\mathcal{F}_m^\alpha$ will change according to

$$\mathcal{F}_k^\alpha \;\to\; \mathcal{F}_k^a \;+\; \mathcal{L}_{(k)}^\alpha[\mathbf{G}_k] \,. \tag{6.83}$$

We can thus eliminate all terms which are in the range of the $\mathcal{L}_{(k)}^\alpha$ operators. If this is done, we say that the system is in *quaternionic normal form*.

In other words, we choose a space complementary to

$$\mathcal{R}_{(k)} \;=\; \bigcup_{\alpha=1}^{3} \mathrm{Ran}(\mathcal{L}_{(k)}^\alpha) \;\subset\; \mathbf{H}_{(k)} \,;$$

if all terms $\mathcal{F}_k^\alpha$ satisfy $\mathcal{F}_k^\alpha \in \mathcal{R}_{(k)}$, for all $k > 1$, the system is in quaternionic normal form.

Remark 6.12 Note we do not have here an equivalent of the Bargmann metric; thus we cannot identify the complementary to the range of $\mathcal{L}_{(k)}^\alpha$ with its kernel, or the kernel of an adjoint operator.[10]

This fact also makes not so easy to implement our general result in concrete computations, or to give an intrinsic characterization of systems in quaternionic normal form.   ⊙

---

[10]We are not aware of any attempt to identify an equivalent of the Bargmann metric in the quaternionic case; in particular, we ignore if there is any real obstacle to its identification. Solving this problem would be a necessary step towards the full development of a quaternionic perturbation theory.

6.4 Normal Forms for Simple Quaternionic Oscillators

We will now apply the general formulas obtained above to the problem of transforming a (nonlinear) simple quaternionic oscillator into normal form.

We will make use of the possibility to operate an SO(3) rotation in the Kahler sphere reduce the quadratic part of the triple of Hamiltonians to the case with $\mathcal{H}_0^2 = \mathcal{H}_0^3 = 0$; moreover as we deal with a quaternionic oscillator we have

$$\mathcal{H}_0^1 = \frac{1}{2} \mu |\mathbf{x}|^2 .$$

We will assume non-degeneracy, i.e. that

$$\mu \neq 0 .$$

Note that in this case the matrix Φ is just $\Phi = \mu I$.

Then our general formulas provide

$$\mathcal{L}_{(k)} : \begin{pmatrix} \mathcal{G}^1 \\ \mathcal{G}^2 \\ \mathcal{G}^3 \end{pmatrix} \rightarrow \begin{pmatrix} \mu x^i Q_\beta^{ij} \partial_j \mathcal{G}_k^\beta \\ 2 \mu \mathcal{G}_k^3 \\ -2 \mu \mathcal{G}_k^2 \end{pmatrix} . \tag{6.84}$$

Thus the net effect of the Poincaré transformation with generator $\mathbf{G}_k$ is to map the terms $\mathcal{F}_k^{\alpha[11]}$ of degree $k + 2$ into new terms of the same degree according to

$$\begin{aligned} \mathcal{F}_k^1 \rightarrow \tilde{\mathcal{F}}_k^1 &= \mathcal{F}_k^1 + \mu x^i Q_\beta^{ij} \partial_j \mathcal{G}_k^\beta , \\ \mathcal{F}_k^2 \rightarrow \tilde{\mathcal{F}}_k^2 &= \mathcal{F}_k^2 - 2 \mu \mathcal{G}_k^3 , \\ \mathcal{F}_k^3 \rightarrow \tilde{\mathcal{F}}_k^3 &= \mathcal{F}_k^3 + 2 \mu \mathcal{G}_k^2 . \end{aligned} \tag{6.85}$$

It is clear that we can always obtain $\tilde{\mathcal{F}}_k^2 = 0 = \tilde{\mathcal{F}}_k^3$ simply by choosing

$$\mathcal{G}_k^2 = -\frac{1}{\mu} \mathcal{F}_k^3 ; \quad \mathcal{G}_k^3 = \frac{1}{\mu} \mathcal{F}_k^2 . \tag{6.86}$$

With this choice, we also obtain

$$\tilde{\mathcal{F}}_k^1 = \mathcal{F}_k^1 + \mu x^i Q_1^{ij} \partial_j \mathcal{G}_k^1 + x^i \left(Q_3^{ij} \partial_j \mathcal{F}_k^2 - x^i Q_2^{ij} \partial_j \mathcal{F}_k^3 \right) + x^i \sum_{\beta=1}^3 \widehat{M}_\beta^{ij} \partial_j \widehat{\mathcal{G}}^\beta$$

if we allow general (Dirac) vector fields as generator; if we only allow generators which are hyperhamiltonian (w.r.t. the underlying hyperkahler structure) the last term—in which we wrote explicitly the sum over β—is absent.

---

[11] Recall that these are the terms obtained after the previous steps in the normalization procedure.

The result of our computation can be expressed also as saying that a simple non-linear quaternionic oscillator with non-degenerate linear part can be put in normal form up to any desired order n_0, and this normal form is a standard Hamiltonian nonlinear oscillator. Note that the terms of order $N > n_0$ will still be in hyperhamiltonian form; in other words we cannot use the standard techniques in Hamiltonian perturbation theory to control these higher order non-normalized terms.

Remark 6.13 It appears that the same discussion can be conducted, with rather obvious modifications, in the multi-component case, i.e. for nonlinear quaternionic oscillators in $\mathbf{R}^{4n}$ and $n > 1$; or at least for the non-resonant case. The effect of resonances among the $\mu_{(k)}$ has not been studied. $\odot$

6.5 Normalization of Quaternionic Oscillators by Quaternionic Oscillators

The situation gets considerably simpler if we limit ourselves to consider generating functions (for the Poincaré transformations) which are themselves Hamiltonians of (nonlinear) quaternionic oscillators.

The reason for this simplification lies in that now $\mathcal{G}_k^\alpha(x) = G_k^\alpha(\rho)$, and hence

$$\partial_i \mathcal{G}_k^a = x^i \frac{dG_k^\alpha}{d\rho}.$$

As all the matrices Q_β, i.e. both the M_α and the $\widehat{M}_\alpha$, are skew-symmetric,

$$x_i \, Q_\beta^{ij} \, \partial_j \mathcal{G}^\beta = \left(x_i \, Q_\beta^{ij} \, x^j \right) \frac{dG^\beta}{d\rho} = 0 \,;$$

that is, all the terms entering in $\widetilde{\mathcal{H}}_{(0)}^\alpha$ vanish.

Note that this hold for any dimension of the system. Note also that considering generators of this form corresponds to considering *normalization by an integrable hyperhamiltonian (or Dirac) flow*.

6.5.1 Simple Quaternionic Oscillator

We start by considering (perturbations of) a simple quaternionic oscillator, so in particular we are in $\mathbf{R}^4$. As usual, we write $\rho = |\mathbf{x}|^2$. We will also write

$$\mathcal{F}^\alpha(x) = F^\alpha(\rho) \,; \quad \mathcal{G}^\alpha(x) = G^\alpha(\rho) \,; \quad \mathcal{H}^\alpha(x) = H^\alpha(\rho) \,.$$

In this case the Poisson bracket terms identically vanish, and H_α turn out to be determined by the equations[12]

$$H_1' = -2 \left(F_2' G_3' - F_3' G_2' \right),$$
$$H_2' = -2 \left(F_3' G_1' - F_1' G_3' \right),$$
$$H_3' = -2 \left(F_1' G_2' - F_2' G_1' \right).$$

These equations do of course provide

$$H_1 = -2 \int \left(F_2' G_3' - F_3' G_2' \right) d\rho,$$

$$H_2 = -2 \int \left(F_3' G_1' - F_1' G_3' \right) d\rho,$$

$$H_3 = -2 \int \left(F_1' G_2' - F_2' G_1' \right) d\rho;$$

We can expand the various function in homogeneous terms, and each of these is by hypothesis just a monomial in ρ (for $n > 1$ things would be more complex):

$$F^a = \sum_k F_k^a = \sum_k \Phi_k^a \rho^{k+1},$$

$$G^a = \sum_k G_k^a = \sum_k \Psi_k^a \rho^{k+1},$$

$$H^a = \sum_k H_k^a = \sum_k \Xi_k^a \rho^{k+1}.$$

Note that here the index k refers to the degree of homogeneity of the various terms $(F_k, G_k, \ldots)$ in ρ, not in x. Thus F_k is homogeneous of degree $k + 1$ in ρ, hence of degree $2k + 2$ in the x, and similarly for the other terms.

Now from the previous equations we have easily that the term coming from F_k^* and G_m^* is

$$\Xi_{k+m}^a = -2 \int (k + m + 2) \, \epsilon_{abc} \, \Phi_k^b \Psi_m^c \, \rho^{k+m} \, d\rho;$$

the integration just gives

$$\Xi_{k+m}^a = -2 \, \frac{k + m + 2}{k + m + 1} \, \epsilon_{abc} \, \Phi_k^b \Psi_m^c \, \rho^{k+m+1}.$$

---

[12] We stress that here we are not (yet) putting the F in standard Hamiltonian form, nor considering the case of quadratic $\mathcal{F}^\alpha$.

In conclusion, we have

$$\Xi^a_{k+m} = -2\,\frac{k+m+2}{k+m+1}\,\epsilon_{abc}\,\Phi^b_k\Psi^c_m$$
$$= K(k,m)\,\epsilon_{abc}\,\Phi^b_k\Psi^c_m := -\langle\Phi_k,\Psi_m\rangle\,. \tag{6.87}$$

In particular, if we think of performing normalizing transformations with generators G_m, the terms F_k with $k < m$ will remain unchanged, while the term F_m will change according to

$$F_m \;\to\; \widetilde{F}_m \;=\; F_m \,+\, (\delta F)_m\,, \tag{6.88}$$

where

$$(\delta F)^a_m \;=\; -\,\langle\Phi_k,\Psi_m\rangle\,. \tag{6.89}$$

Now we would like to cancel the terms F_m ($m > 0$) by choosing appropriately the G_m; this appropriate choice would be given by the solutions to the *homological equation* (to be derived through the usual Poincaré procedure), which in this framework reads

$$\Xi_m \;=\; \Phi_m \;=\; \langle\Phi_0,\Psi_m\rangle\,; \tag{6.90}$$

this should be seen as an equation for the Ψ^a_m.

Writing this in components, we get a vector equation of the form

$$\Xi \;=\; \Gamma\,\Psi\,; \tag{6.91}$$

more precisely we have (the scalar function $K(k,m)$ is defined in (6.87))

$$\begin{pmatrix}\Xi^1_m\\\Xi^2_m\\\Xi^3_m\end{pmatrix} \;=\; K(k,m)\begin{pmatrix}0 & -\Phi^3_0 & \Phi^2_0\\\Phi^3_0 & 0 & -\Phi^1_0\\-\Phi^2_0 & \Phi^1_0 & 0\end{pmatrix}\begin{pmatrix}\Psi^1_m\\\Psi^2_m\\\Psi^3_m\end{pmatrix}\,. \tag{6.92}$$

It is clear that $\det(\Gamma) = 0$, so we cannot eliminate all terms in F_m, but only those in the range of $\Gamma \approx \mathcal{L}$.

Note that in this case the matrix representation Γ of the homological operator is the same (and anyway of the same dimension) for all the degrees. It will be invertible on two-dimensional subspaces. In particular, we can eliminate terms in the same two-dimensional subspace at all degrees.

We summarize our discussion in a formal statement.

Lemma 6.3 *By a suitable sequence of Poincaré transformations corresponding to the flow of (integrable) hyperhamiltonian vector fields, any (integrable) hyperhamiltonian vector field*

$$X \;=\; f^i\partial_i\,, \qquad f^i \;=\; \sum_\alpha M^{ij}_\alpha\partial_j F^\alpha \tag{6.93}$$

can be brought, in the vicinity of a fixed point, to the form

$$X = \lambda^i \partial_i \tag{6.94}$$

where

$$\lambda^i = \sum_\alpha M_\alpha^{ij} \partial_j \Lambda^\alpha , \tag{6.95}$$

with $\Lambda^2 = F_0^2$, $\Lambda^3 = F_0^3$.

Remark 6.14 If now we use the possibility of taking the linear part to Hamiltonian by a rotation in the ω_a (i.e. by an SO(3) rotation in the Kahler sphere), see also Remark 6.3, the linear part of the system is Hamiltonian, and we can always set $F_0^2 = F_0^3 = 0$. By Lemma 6.3 the perturbation is also (locally, as the normal form approach is intrinsically local) Hamiltonian up to any desired order n_0. This is of course the same phenomenon mentioned above, see Sect. 6.4.   ⊙

Remark 6.15 To obtain the form stated in Lemma 6.3, in particular (6.95), we can choose at any step

$$\Psi_m^2 = \frac{\Phi_m^3 + \Psi_m^1 K \Phi_0^2}{K \Phi_0^1} , \quad \Psi_m^3 = \frac{\Phi_m^3 \Phi_0^3 + \Psi_m^1 K \Phi_0^2 \Phi_0^3 + \Phi_m^1 \varphi_1}{K \Phi_0^2 \Phi_0^1} ; \tag{6.96}$$

this of course applies provided Φ_0^1 and Φ_0^2 are both nonzero.   ⊙

6.6 Normalization by Dirac Vector Fields

We have considered normalization of a hyperhamiltonian vector field by hyperhamiltonian transformations; or, in the language of Lie transforms, by conjugation with another hyperhamiltonian vector field. As mentioned above, see Remark 6.5, one could consider also normalization by more general transformations, i.e. those generated by Dirac vector fields.

Our notation, in particular the use of Q_α to denote the hyperkahler matrices associated to the normalizing vector fields, allows to adapt the results obtained in Sects. 6.4 and 6.5 without the need to repeat all the computations.

As usual, we will first consider the four-dimensional (i.e. $n = 1$) case and then the general $4n$-dimensional one.

6.6.1 Simple Quaternionic Oscillator

In the case $n = 1$ we just have to deal with two standard hyperkahler structures; that is the defining one **J**—say, for definiteness and as above, the positively oriented one, corresponding to matrices M_α; and the dual one $\widehat{\mathbf{J}}$, i.e. with this choice the negatively oriented one with associated matrices $\widehat{M}_\alpha$.

6.6.1.1 Normalization by Dual Hyperhamiltonian Vector Fields

If we consider normalizing transformations generated by vector fields which are hyperhamiltonian w.r.t. the $\widehat{\mathbf{J}}$ structure, this just amounts to having three Q_α matrices, with

$$Q_\alpha = \widehat{M}_\alpha .$$

All of our computations remain correct, but there is a difference. That is, we know that

$$\left[M_\alpha , \widehat{M}_\beta \right] = 0 \quad \forall \alpha, \beta . \tag{6.97}$$

As a consequence, all the commutator terms will vanish. This means on the one side that normalization by oscillators is impossible (recall the Poisson bracket terms are identically zero when the generating functions are also of oscillator type); and on the other that normalization with general generating functions **G** acts only by the "usual" Poisson bracket terms. Note also that this acts component-wise in the sense of the basis in **Q** (i.e. of the hyperkahler index α).

6.6.1.2 Normalization by General Dirac Hyperhamiltonian Vector Fields

Let us consider the case where we deal with normalization by general Dirac vector field, i.e. with normalizing vector fields of the form

$$Y = g^i(x)\, \partial_i ; \quad g^i = M_\alpha^{ij}\, \partial_j \mathcal{H}^\alpha + \widehat{M}_\alpha^{ij}\, \partial_j \widehat{\mathcal{H}}^\alpha . \tag{6.98}$$

In this case the Eq. (6.85), describing the effect of general normalizing transformations, are replaced by

$$\begin{aligned}
\mathcal{F}_k^1 \to \widetilde{\mathcal{F}}_k^1 &= \mathcal{F}_k^1 + \mu x^i M_\beta^{ij}\, \partial_j \mathcal{G}_k^\beta + \mu x^i \widehat{M}_\beta^{ij}\, \partial_j \widehat{\mathcal{G}}_k^\beta , \\
\mathcal{F}_k^2 \to \widetilde{\mathcal{F}}_k^2 &= \mathcal{F}_k^2 - 2\mu\, \mathcal{G}_k^3 , \\
\mathcal{F}_k^3 \to \widetilde{\mathcal{F}}_k^3 &= \mathcal{F}_k^3 + 2\mu\, \mathcal{G}_k^2 .
\end{aligned} \tag{6.99}$$

That is, we obtain the same as before with an additional term (the one with the $\widehat{\mathcal{G}}_k^\beta$ functions); this gives, at least in principles, the possibility of eliminating terms which cannot be reached by normalizing only with a hyperhamiltonian vector field.

6.6.2 General Quaternionic Oscillator

In the case of systems in $(\mathbf{R}^{4n}, \delta)$ with a given hyperkahler structure $\mathbf{J}$ (let us say for definiteness the one which is positively oriented in each of the $\mathbf{R}^4$ irreducible hyperkahler subspaces) we have not just one dual hyperkahler structure, but several ones. In fact, as we have discussed in Chap.1, we have $(2^n - 1)$ structures which are dual to the given one.

6.6.2.1 Normalization by Dual Hyperhamiltonian Vector Fields

If we consider normalizing vector fields which are associated to just the dual structure which is orientation-reversed in each of the $\mathbf{R}^4$ irreducible hyperkahler subspaces,

$$Y = \left(\widehat{M}_\alpha^{ij} \, \partial_j \widehat{\mathcal{H}}^\alpha \right) \partial_i \,,$$

the same considerations presented in the $n = 1$ case apply. That is, now all the commutator terms vanish, and we are just left with the action by the Poisson terms.

In particular, (6.85)—referring to the case where the linear part was set in standard Hamiltonian form—reads now (with s the block index)

$$\begin{aligned}
\mathcal{F}_k^1 \to \widetilde{\mathcal{F}}_k^1 &= \mathcal{F}_k^1 + \mu^{(s)} x^i (M_\beta)_{(s)}^{ij} \partial_j \mathcal{G}_k^\beta \,, \\
\mathcal{F}_k^2 \to \widetilde{\mathcal{F}}_k^2 &= \mathcal{F}_k^2 \,, \\
\mathcal{F}_k^3 \to \widetilde{\mathcal{F}}_k^3 &= \mathcal{F}_k^3 \,.
\end{aligned} \qquad (6.100)$$

That is, only normalization of the $\mathcal{F}^1$ Hamiltonian occurs. This is entirely analogous to the standard Birkhoff normalization.

6.6.2.2 Normalization by Strictly Dirac Hyperhamiltonian Vector Fields

We recall that a strictly Dirac vector field for a given hyperkahler structure $\mathbf{J}$ is one which is the superposition of two vector fields, one which is hyperhamiltonian w.r.t. $\mathbf{J}$ and one which is hyperhamiltonian w.r.t. the dual structure $\widehat{\mathbf{J}}$ which is orientation-reversed in each of the $\mathbf{R}^4$ irreducible hyperkahler subspaces,

$$Y = \left[\left(M_\alpha^{ij} \, \partial_j \mathcal{H}^\alpha \right) + \left(\widehat{M}_\alpha^{ij} \, \partial_j \widehat{\mathcal{H}}^\alpha \right) \right] \partial_i \,.$$

In this case also the discussion of the Dirac $n = 1$ case applies, and we get similarly to

$$
\begin{aligned}
\mathcal{F}_k^1 \to \widetilde{\mathcal{F}}_k^1 &= \mathcal{F}_k^1 + \mu^{(s)} x^i (M_\beta)^{ij}_{(s)} \partial_j \mathcal{G}_k^\beta + \mu^{(s)} x^i (\widehat{M}_\beta)^{ij}_{(s)} \partial_j \widehat{\mathcal{G}}_k^\beta \,, \\
\mathcal{F}_k^2 \to \widetilde{\mathcal{F}}_k^2 &= \mathcal{F}_k^2 - 2\mu\, \mathcal{G}_k^3 \,, \\
\mathcal{F}_k^3 \to \widetilde{\mathcal{F}}_k^3 &= \mathcal{F}_k^3 + 2\mu\, \mathcal{G}_k^2 \,,
\end{aligned}
\tag{6.101}
$$

where now (with an abuse of notation) M_α and $\widehat{M}_\alpha$ should be meant as the $4n$-dimensional block-diagonal matrices having the four-dimensional M_α and $\widehat{M}_\alpha$ matrices as diagonal blocks.

6.6.2.3 Normalization by General Dirac Hyperhamiltonian Vector Fields

The situation is apparently more complex for general Dirac fields, in that several hyperkahler structure enter in the play, and for each of these and each block we have positive or negative orientation; and we should have 2^n terms in Y and hence in (6.99) and (6.101).

However, note that (6.99) are linear in the $\mathcal{G}^\alpha$, $\widehat{\mathcal{G}}^\alpha$ functions; thus all the 2^{n-1} terms in which a $(M_\alpha)_{(s)}$ matrix appears (note that each of these would have its own $\mathcal{G}^\alpha$ function) can be summed up, and similarly for terms with a $(\widehat{M}_\alpha)_{(s)}$ matrix.

So in the end it suffices to consider strictly Dirac vector fields, obtaining the same range of transformations.

Chapter 7
Physical Applications

We will discuss in this chapter two applications of the theory to physically interesting cases, dealing with dynamics of particles with spin 1/2 in a magnetic field, i.e. the Pauli and the Dirac equations; and an explicit computation of the hyperkahler structure of the Taub-NUT metrics.

While the Pauli equation corresponds to a hyperhamiltonian flow, it turns out that the hyperhamiltonian description of the Dirac equation is in terms of two commuting hyperhamiltonian flows. In this framework one can use a factorization principle, discussed here (which is a special case of a general phenomenon studied by Walcher [144]) and provide an explicit description of the resulting flow. On the other hand, by applying the familiar Foldy-Wouthuysen and Cini-Touschek transformations (and the recently introduced Mulligan transformation) which separate—in suitable limits—the Dirac equation into two equations, each of these turns out to be described by a single hyperhamiltonian flow. Thus the hyperhamiltonian construction is able to describe the fundamental dynamics for particles with spin.

As it should be obvious at this point, hyperkahler structures are naturally related to *spin structures*. It is thus entirely natural, from the physical point of view, to investigate if and to which extent hyperhamiltonian dynamics is relevant to the Physics of systems with spin.

We will show in this Chapter how hyperhamiltonian dynamics applies to physical equations describing the evolution of the spin degrees of freedom of particles, i.e. to the Pauli and Dirac equations.

While the Majorana-Weyl equation is immediately set in standard hyperhamiltonian form (due to a degeneration of the Majorana-Weyl equation, corresponding to $m = 0$; other degenerated Dirac equations can also be reduced to hyperhamiltonian form), the full Dirac equation defines a flow which is the sum of two hyperhamiltonian ones, associated to conjugate hyperkahler structures.

© Springer International Publishing AG 2017

G. Gaeta and M.A. Rodríguez, *Lectures on Hyperhamiltonian Dynamics and Physical Applications*, Mathematical Physics Studies, DOI 10.1007/978-3-319-54358-1_7

It is known that the free Dirac equation in $\mathbf{C}^4$ can be separated in two spinor $\mathbf{C}^2$ equations by means of non-local transformations, such as the Foldy-Wouthuysen (FW) one [57] (appropriate in the non-relativistic limit) or the Cini-Touschek one [48] (appropriate in the ultra-relativistic limit); the latter has been recently reconsidered by Mulligan [122]. In all these cases, of course, separation does not occur in the presence of an electromagnetic field, albeit it can be obtained perturbatively up to a given order in a suitable expansion parameter (e.g., in $\varepsilon = \hbar/mc^2$ for the FW case). It turns out that in this case the full equation can not be expressed as a hyperhamiltonian flow; but it is still possible to express the equations in terms of two hyperhamiltonian vector fields as above.

7.1 The Pauli Equation

The non-relativistic evolution equation for particles with spin one-half is provided by the Pauli equation, see e.g. [110].[1]

The 2-component wave function of a spin 1/2 charged particle with fixed momentum $\mathbf{p}$ in an electromagnetic field[2] satisfies the Pauli equation

$$i\,\hbar\,\partial_t\varphi \;=\; \left[\left(\frac{1}{2m}(\mathbf{p} - e\mathbf{A})^2 + e\Phi\right)\sigma_0 - \frac{e\hbar}{2m}\sigma\mathbf{B}\right]\varphi\,, \tag{7.1}$$

where the fields Φ and $\mathbf{B}$ depend only on time.

We will consider in the sequel the standard $\mathbf{Q}$-structures Y and $\widehat{Y}$ corresponding to the positive (1.23) and negative orientations (1.25), respectively.

The hyperhamiltonian equations of motion are given by

$$\dot{x}^1 = \partial_2\mathcal{H}^1 + \partial_4\mathcal{H}^2 + \partial_3\mathcal{H}^3\;;\;\dot{x}^2 = -\partial_1\mathcal{H}^1 + \partial_3\mathcal{H}^2 - \partial_4\mathcal{H}^3;$$
$$\dot{x}^3 = \partial_4\mathcal{H}^1 - \partial_2\mathcal{H}^2 - \partial_1\mathcal{H}^3\;;\;\dot{x}^4 = -\partial_3\mathcal{H}^1 - \partial_1\mathcal{H}^2 + \partial_2\mathcal{H}^3\,. \tag{7.2}$$

in the positive oriented case, and

$$\dot{x}^1 = \partial_3\mathcal{H}^1 - \partial_4\mathcal{H}^2 - \partial_2\mathcal{H}^3\;\;;\;\dot{x}^2 = \partial_4\mathcal{H}^1 + \partial_3\mathcal{H}^2 + \partial_1\mathcal{H}^3\;;$$
$$\dot{x}^3 = -\partial_1\mathcal{H}^1 - \partial_2\mathcal{H}^2 + \partial_4\mathcal{H}^3\;;\;\dot{x}^4 = -\partial_2\mathcal{H}^1 + \partial_1\mathcal{H}^2 - \partial_3\mathcal{H}^3\,. \tag{7.3}$$

in the negative one.

The space $\mathbf{R}^4$ in which the Y, $\widehat{Y}$ act is of course isomorphic to $\mathbf{C}^2$, and conversely the space $\mathbf{C}^2$ in which Pauli matrices act is isomorphic to $\mathbf{R}^4$; this isomorphism is however not unique, and depends on the choice of a basis in $\mathbf{R}^4$. Thus our way to express equations in which the Pauli matrices appear in terms of the Y, $\widehat{Y}$ matrices,

---

[1]The hyperhamiltonian framework for this equation has been considered elsewhere [60, 61] in the simplified setting of no electric field.

[2]This will depend only on time, as it also follows from the fact we are considering the momentum representation for the particle, i.e. its position is completely undetermined.

will depend on this choice. As in practice they always appear as $i\sigma_\mu$, we are interested in the expression of these quantities in terms of our quaternionic (that is, hyperkahler) matrices.

With the basis

$$\widehat{v}_1 = \begin{pmatrix} 1 \\ 0 \end{pmatrix}, \ \widehat{v}_2 = \begin{pmatrix} i \\ 0 \end{pmatrix}, \ \widehat{v}_3 = \begin{pmatrix} 0 \\ 1 \end{pmatrix}, \ \widehat{v}_4 = \begin{pmatrix} 0 \\ i \end{pmatrix} \tag{7.4}$$

we have

$$i\sigma_0 \simeq -Y_1, \quad i\sigma_1 \simeq \widehat{Y}_2, \quad i\sigma_2 \simeq \widehat{Y}_1, \quad i\sigma_3 \simeq \widehat{Y}_3. \tag{7.5}$$

Choosing other bases would give different correspondences. For instance, if

$$v_1 = \begin{pmatrix} 1 \\ 0 \end{pmatrix}, \ v_2 = \begin{pmatrix} 0 \\ 1 \end{pmatrix}, \ v_3 = \begin{pmatrix} i \\ 0 \end{pmatrix}, \ v_4 = \begin{pmatrix} 0 \\ i \end{pmatrix} \tag{7.6}$$

we have:

$$i\sigma_0 \simeq -\widehat{Y}_1, \quad i\sigma_1 \simeq -Y_2, \quad i\sigma_2 \simeq Y_1, \quad i\sigma_3 \simeq -Y_3. \tag{7.7}$$

These correspondences will be of use in the following; we will work mainly with the basis (7.4) and the correspondence (7.5).

Choosing still other bases would give a correspondence equivalent—via an SU(2) automorphism—to either one of (7.5) or (7.7), depending on orientation.

We note that

$$Y_1 = \begin{pmatrix} i\sigma_2 & 0 \\ 0 & i\sigma_2 \end{pmatrix}, \quad Y_2 = \begin{pmatrix} 0 & \sigma_1 \\ -\sigma_1 & 0 \end{pmatrix}, \quad Y_3 = \begin{pmatrix} 0 & \sigma_3 \\ -\sigma_3 & 0 \end{pmatrix}; \tag{7.8}$$

$$\widehat{Y}_1 = \begin{pmatrix} 0 & \sigma_0 \\ -\sigma_0 & 0 \end{pmatrix}, \quad \widehat{Y}_2 = \begin{pmatrix} 0 & -i\sigma_2 \\ -i\sigma_2 & 0 \end{pmatrix}, \quad \widehat{Y}_3 = \begin{pmatrix} -i\sigma_2 & 0 \\ 0 & i\sigma_2 \end{pmatrix}. \tag{7.9}$$

(Obviously by multiplying the three matrices of a structure by the same σ_α matrix we get an equivalent one). Actually, this notation may be misleading, as it refers to matrices acting in $\mathbf{C}^4$, while the Y_α and $\widehat{Y}_\alpha$ matrices should act—representing quaternionic operations—in $\mathbf{H}^1 = \mathbf{R}^4$.

Note that the symplectic forms corresponding to the hyperkahler and metric structures: ω_α (respectively, the $\widehat{\omega}_\alpha$), are a basis for the space of self-dual (respectively, anti-self-dual) two-forms in $\mathbf{R}^4$; thus we say that (7.2) describes self-dual hyper-hamiltonian dynamics, and (7.3) describes the anti-self-dual one.

Using the basis (7.4), that is

$$\varphi = \begin{pmatrix} \chi \\ \zeta \end{pmatrix}, \quad \widehat{\Theta} = \begin{pmatrix} \mathrm{Re}\,\chi \\ \mathrm{Im}\,\chi \\ \mathrm{Re}\,\zeta \\ \mathrm{Im}\,\zeta \end{pmatrix} = \begin{pmatrix} \chi_+ \\ \chi_- \\ \zeta_+ \\ \zeta_- \end{pmatrix}. \tag{7.10}$$

and the correspondence (7.5), we immediately get

$$\partial_t \widehat{\Theta} = \left[K\, Y_1 + \frac{e}{2m}\, (B_y \widehat{Y}_1 + B_x \widehat{Y}_2 + B_z \widehat{Y}_3) \right] \widehat{\Theta} . \tag{7.11}$$

where

$$K = \frac{1}{\hbar} \left(\frac{1}{2m}\, (\mathbf{p} - e\mathbf{A})^2 + e\,\Phi \right), \tag{7.12}$$

In other words, the Pauli equation is written as

$$\partial_t \widehat{\Theta} = \widehat{H}\, \widehat{\Theta} \tag{7.13}$$

with $\widehat{H}$ given by

$$\widehat{H} = K\, Y_1 + \frac{e}{2m}\, (B_y \widehat{Y}_1 + B_x \widehat{Y}_2 + B_z \widehat{Y}_3). \tag{7.14}$$

This is in hyperhamiltonian form if $K = 0$; in general the flow is described by the sum of two hyperhamiltonian flows, one with respect to the $\{\widehat{Y}_i\}$ structures and one with respect to the $\{Y_i\}$ structures—albeit only Y_1 actually appears.

For $K = 0$ the equation reduces to (here $\partial_i = \partial_{\Theta_i}$)

$$\begin{aligned}
\partial_t \chi_+ &= \partial_3 \widehat{\mathcal{H}}^1 - \partial_4 \widehat{\mathcal{H}}^2 - \partial_2 \widehat{\mathcal{H}}^3 \\
\partial_t \chi_- &= -\partial_1 \widehat{\mathcal{H}}^1 - \partial_2 \widehat{\mathcal{H}}^2 + \partial_4 \widehat{\mathcal{H}}^3 \\
\partial_t \zeta_+ &= \partial_4 \widehat{\mathcal{H}}^1 + \partial_3 \widehat{\mathcal{H}}^2 + \partial_1 \widehat{\mathcal{H}}^3 \\
\partial_t \zeta_- &= -\partial_2 \widehat{\mathcal{H}}^1 + \partial_1 \widehat{\mathcal{H}}^2 - \partial_3 \widehat{\mathcal{H}}^3
\end{aligned} \tag{7.15}$$

where

$$\widehat{\mathcal{H}}^1 = \frac{e|\widehat{\Theta}|^2}{4m}\, B_y , \quad \widehat{\mathcal{H}}^2 = \frac{e|\widehat{\Theta}|^2}{4m}\, B_x , \quad \widehat{\mathcal{H}}^3 = \frac{e|\widehat{\Theta}|^2}{4m}\, B_z . \tag{7.16}$$

It should be noted that the order in which the $\chi_\pm, \zeta_\pm$ enter in $\widehat{\Theta}$ was chosen arbitrarily; it is interesting to observe what happens choosing a different order, i.e. defining

$$\Theta = \begin{pmatrix} \chi_+ \\ \zeta_+ \\ \chi_- \\ \zeta_- \end{pmatrix} . \tag{7.17}$$

that is, using the second basis (7.6) and the correspondence (7.7). In this case (7.1) reads

$$\partial_t \Theta = \left[K\, \widehat{Y}_1 + \frac{e}{2m}\, (B_y Y_1 - B_x Y_2 - B_z Y_3) \right] \Theta . \tag{7.18}$$

Thus, in this representation the Pauli equation reads

$$\partial_t \Theta = H \Theta \tag{7.19}$$

with H given by

$$H = K \widehat{Y}_1 + \frac{e}{2m} (B_y Y_1 - B_x Y_2 - B_z Y_3). \tag{7.20}$$

That is, we have a role reversal of the two standard hyperhamiltonian structures: again we get an equation in hyperhamiltonian form (this time with the Y_i rather than the $\widehat{Y}_i$ complex structures) if $K = 0$; in general the flow is described by the sum of two hyperhamiltonian flows.

For $K = 0$ the equation reduces to

$$\begin{aligned}
\partial_t \chi_+ &= \partial_2 \mathcal{H}^1 + \partial_4 \mathcal{H}^2 + \partial_3 \mathcal{H}^3 , \\
\partial_t \zeta_+ &= \partial_4 \mathcal{H}^1 - \partial_2 \mathcal{H}^2 - \partial_1 \mathcal{H}^3 , \\
\partial_t \chi_- &= -\partial_1 \mathcal{H}^1 + \partial_3 \mathcal{H}^2 - \partial_4 \mathcal{H}^3 , \\
\partial_t \zeta_- &= -\partial_3 \mathcal{H}^1 - \partial_1 \mathcal{H}^2 + \partial_2 \mathcal{H}^3
\end{aligned} \tag{7.21}$$

where

$$\mathcal{H}^1 = \frac{e|\Theta|^2}{4m} B_y , \quad \mathcal{H}^2 = -\frac{e|\Theta|^2}{4m} B_x , \quad \mathcal{H}^3 = -\frac{e|\Theta|^2}{4m} B_z . \tag{7.22}$$

The possibility of expressing the equations using the two standard hyperkahler structures is related to the following fact: if

$$P = \begin{pmatrix} 1 & 0 & 0 & 0 \\ 0 & 0 & 1 & 0 \\ 0 & 1 & 0 & 0 \\ 0 & 0 & 0 & 1 \end{pmatrix} , \tag{7.23}$$

then P is an intertwining operator [105] for the two standard *real* representations of su(2) and hence for the two standard hyperhamiltonian structures:

$$P (\alpha Y_1 + \beta Y_2 + \gamma Y_3) P^{-1} = \alpha \widehat{Y}_1 - \beta \widehat{Y}_2 - \gamma \widehat{Y}_3 . \tag{7.24}$$

7.2 The Dirac Equation

The proper formalism to discuss relativistic particles with spin 1/2 is provided by the Dirac equation [22, 98, 110, 117]; in the massless case this takes the form of the Majorana-Weyl equation.[3]

---

[3]Notice that these are field PDEs rather than ODEs, so that hyperhamiltonian dynamics can deal only with finite dimensional reductions of them. This remark would call for the extension of hyperhamiltonian dynamics to a hyperhamiltonian field theory; a task which lies beyond the limits of the present discussion.

We follow the convention and notation of [22]; in particular, the metric is given by $(+1, -1, -1, -1)$. Greek indices run from 0 to 3, latin indices from 1 to 3.

7.2.1 Dirac Equation: Generalities

Let us start from the Dirac equation written in terms of the γ matrices,

$$(\gamma^\mu p_\mu - mc)\,\psi = 0 ; \tag{7.25}$$

here ψ is a bi-spinor (four complex components), and we work in the momentum representation; hence $\mathbf{p}$ can be considered as a constant. In fact, in what follows we will consider the equation

$$[i\hbar\gamma^0\,\partial_t - c(\boldsymbol{\gamma}\cdot\mathbf{p}) - mc^2]\,\psi = 0 \tag{7.26}$$

with ψ depending only on t. This can also be written as

$$i\hbar\,\partial_t\psi = \gamma^0[c(\boldsymbol{\gamma}\cdot\mathbf{p}) + mc^2]\,\psi = [c(\mathbf{p}\cdot\boldsymbol{\alpha}) + mc^2\beta]\,\psi \tag{7.27}$$

We will consider two different representations for the Dirac equation: the standard and the spinorial representations.

Before presenting detailed computations, let us present two rather obvious remarks.

• There is no hope to find a representation of the full Dirac equation in terms of only one hyperkahler structure, essentially because γ matrices are four and Y (or $\widehat{Y}$) matrices are only three.

• We stress that the choice of one or another hyperkahler structure depends on the order one chooses to write the equations in matrix form (i.e. on the orientation of the spin space). Anyway, for the full Dirac equation we will always get the whole set of complex structures of one of the hyperkahler structures and only one matrix of the other. This other matrix depends on the representation used to write the γ matrices (it is not necessarily associated to the mass term). In fact it is related to the essentially non-diagonal pattern of γ matrices in any 4×4 representation (or, in other words, to the irreducibility property of the representation of the Clifford algebra).

7.2.2 Dirac Equation: Standard Representation

If we choose the standard representation of the γ matrices, we have

$$\beta = \begin{pmatrix} \sigma_0 & 0 \\ 0 & -\sigma_0 \end{pmatrix}, \quad \alpha = \begin{pmatrix} 0 & \boldsymbol{\sigma} \\ \boldsymbol{\sigma} & 0 \end{pmatrix} . \tag{7.28}$$

With this we write the Dirac equation for a spin 1/2 point particle in interaction with an external electromagnetic field $A = (\Phi, \mathbf{A})$ as

$$i\hbar \frac{\partial \Psi}{\partial t} = \left[c\alpha \cdot \left(\mathbf{p} - \frac{e}{c}\mathbf{A} \right) + mc^2\beta + e\Phi I \right] \Psi . \tag{7.29}$$

Here e is the charge of the particle, and we write $\pi := \mathbf{p} - \frac{e}{c}\mathbf{A}$. Hence (7.29) reads

$$\partial_t \Psi = -\frac{i}{\hbar} \left[c\alpha \cdot \pi + (mc^2)\beta + e\Phi \right] \Psi . \tag{7.30}$$

Using (7.28) and $\Psi = (\Psi_+, \Psi_-)^T$, this is in turn rewritten as

$$\partial_t \begin{pmatrix} \Psi_+ \\ \Psi_- \end{pmatrix} = -\frac{i}{\hbar} \begin{pmatrix} (mc^2 + e\Phi)\,\sigma_0 & c\,\boldsymbol{\mathrm{œ}} \cdot \pi \\ c\,\boldsymbol{\mathrm{œ}} \cdot \pi & -(mc^2 - e\Phi)\,\sigma_0 \end{pmatrix} \begin{pmatrix} \Psi_+ \\ \Psi_- \end{pmatrix} . \tag{7.31}$$

Recalling now (7.5), writing for short

$$\widehat{\mathbf{K}} := \widehat{Y}_2 \pi^1 + \widehat{Y}_1 \pi^2 + \widehat{Y}_3 \pi^3 , \tag{7.32}$$

and understanding that the complex quantities $\Psi_\pm \in \mathbf{C}^2$ are represented by four-dimensional real vectors

$$\Psi_\pm = \left(\mathrm{Re}(\Psi_\pm^1), \mathrm{Im}(\Psi_\pm^1), \mathrm{Re}(\Psi_\pm^2), \mathrm{Im}(\Psi_\pm^2) \right)^T$$

in which we are using (7.4), so that σ matrices will be represented according to (7.5), we can therefore rewrite (7.31) in the form

$$\partial_t \begin{pmatrix} \Psi_+ \\ \Psi_- \end{pmatrix} = \begin{pmatrix} [(mc^2 + e\Phi)/\hbar]\, Y_1 & -(c/\hbar)\,\widehat{\mathbf{K}} \\ -(c/\hbar)\,\widehat{\mathbf{K}} & -[(mc^2 - e\Phi)/\hbar]\, Y_1 \end{pmatrix} \begin{pmatrix} \Psi_+ \\ \Psi_- \end{pmatrix} . \tag{7.33}$$

Thus the flow of (7.34), i.e. of the general Dirac equation, is the composition of two hyperhamiltonian flows; these commute, as noted in Sect. 5.4. The whole discussion of Sect. 5.4—in particular, the factorization principle—does therefore apply to the full Dirac equation (in the standard representation).

It is maybe convenient to pass to variables $\xi_\pm = \Psi_+ \pm \Psi_- \in \mathbf{R}^4$. In terms of these we have

$$\begin{aligned} \hbar\dot{\xi}_+ &= \left[(e\Phi)\, Y_1 - c\,\widehat{\mathbf{K}} \right] \xi_+ + (mc^2)\, Y_1\, \xi_- \\ \hbar\dot{\xi}_- &= \left[(e\Phi)\, Y_1 - c\,\widehat{\mathbf{K}} \right] \xi_- + (mc^2)\, Y_1\, \xi_+ . \end{aligned} \tag{7.34}$$

As well known, we can actually always set $\Phi = 0$ by a gauge transformation (this is the Lorentz gauge), so that we can always reduce the full Dirac equation to the form

$$\begin{aligned} \hbar\dot{\xi}_+ &= -c\,\widehat{\mathbf{K}}\, \xi_+ + (mc^2)\, Y_1\, \xi_- \\ \hbar\dot{\xi}_- &= c\,\widehat{\mathbf{K}}\, \xi_- + (mc^2)\, Y_1\, \xi_+ . \end{aligned} \tag{7.35}$$

Needless to say, this is again the sum of two commuting hyperhamiltonian flows.

It is immediate to note that the Eq. (7.35) for $m = 0$—also known as the *Majorana-Weyl equation*—is therefore written in hyperhamiltonian form with the use of only *one* standard hyperkahler structure.

7.2.3 Dirac Equation: Spinor Representation

Let us now choose the spinor representation for the γ matrices. Then

$$\beta = \begin{pmatrix} 0 & \sigma_0 \\ \sigma_0 & 0 \end{pmatrix} , \quad \alpha = \begin{pmatrix} \sigma & 0 \\ 0 & -\sigma \end{pmatrix} . \tag{7.36}$$

We will now, for the sake of brevity, just work in the Lorentz gauge.

We write again the wave function as $\Psi = (\Psi_+, \Psi_-)^T$, and use $\widehat{\mathbf{K}}$ also as above. Proceeding as before, we get the Dirac equation as

$$\partial_t \begin{pmatrix} \Psi_+ \\ \Psi_- \end{pmatrix} = \begin{pmatrix} -(c/\hbar)\,\widehat{\mathbf{K}} & \left[(mc^2)/\hbar\right] Y_1 \\ \left[(mc^2)/\hbar\right] Y_1 & (c/\hbar)\,\widehat{\mathbf{K}} \end{pmatrix} \begin{pmatrix} \Psi_+ \\ \Psi_- \end{pmatrix} . \tag{7.37}$$

It is thus clear that again both structures appear, although one of them only in the mass term (recall we are working in the Lorentz gauge). Thus the general Dirac equation in spinor representation is written as the sum of two commuting hyperhamiltonian flows, and the discussion of Sect. 7.3 applies.[4]

One can pass from the standard to the spinorial representation by the unitary transformation

$$U = \frac{1}{\sqrt{2}} \begin{pmatrix} \sigma^0 & \sigma^0 \\ \sigma^0 & -\sigma^0 \end{pmatrix} ; \tag{7.38}$$

indeed, $\gamma_{\mathrm{st}}^\mu = U^+(\gamma_{\mathrm{sp}}^\mu)U$. Note that the transformation U is real.

The two standard hyperhamiltonian structures enter in the Dirac equation in an asymmetric way; however—as was the case also for the Pauli equation—their role is reversed by just changing the representation of the relevant matrices and vectors. This applies both to the present and the previous subsection.

7.2.4 Separating the Dirac Equation

One would like to separate the Dirac equation into two equations, e.g. for the positive and negative energy states. This is in general impossible, but can be done (via a recursive procedure) up to some given order in a certain expansion parame-

---

[4]Again, the Majorana-Weyl equation would be written in simple hyperhamiltonian form.

ter. There exists indeed *"a systematic procedure developed by Foldy and Wouthuysen, namely, a canonical transformation which decouples the Dirac equation into two two-component equations: one reduces to the Pauli description in the nonrelativistic limit; the other describes the negative-energy states"* (see [22], vol. I page 46).

In this way one obtains, in the general case, an equation which represents a perturbation of a pair of separate equations, i.e. the coupling term between Ψ_+ and Ψ_- is of order ε^k, with ε a suitable perturbation parameter.

A similar procedure, valid in the ultra-relativistic limit, was developed by Cini and Touschek [48] (see also the recent extension by Mulligan [122]). These are considered below.

7.2.5 Unitary Transformations and the Dirac Equation

Once we operate with a unitary transformation $\Psi \to \Psi' = e^{iS}\Psi$ with generator S, where $\Psi = (\Psi_+, \Psi_-)^T$, the Dirac equation $i\hbar\Psi_t = H_0\Psi$ (with H_0 the Dirac Hamiltonian) is transformed into

$$\Psi'_t = H_s \Psi' \tag{7.39}$$

with H_s the transformed Dirac Hamiltonian,

$$H_s = e^{-iS} H_0 e^{iS} + e^{-iS}\partial_t(e^{iS}). \tag{7.40}$$

With such a transformation one can transform the Dirac equation into a different form, which may be more convenient in a given limit.

Here we will work in the free case for ease of discussion; we refer to [98, 117] for the general case.

The matrices Y_i and $\widehat{Y}_i$, as well as the β, α^i, will be as above. We use the standard representation for γ-matrices (we will not change this representation in the sequel, except for some comments on the spinor representation); the Dirac equation is written as

$$\hbar \begin{pmatrix} \dot{\psi}_+ \\ \dot{\psi}_- \end{pmatrix} = H \begin{pmatrix} \psi_+ \\ \psi_- \end{pmatrix} \tag{7.41}$$

where H is given by

$$H = \begin{pmatrix} mc^2\, Y_1 & -c\,\mathbf{K} \\ -c\,\mathbf{K} & -mc^2\, Y_1 \end{pmatrix} \tag{7.42}$$

with $\mathbf{K} = \widehat{Y}_2 p^1 + \widehat{Y}_1 p^2 + \widehat{Y}_3 p^3$.

7.2.6   The Foldy-Wouthuysen Transformation

The Foldy-Wouthuysen transformation is given by

$$U_{FW} = \sqrt{\frac{E + mc^2}{2E}} I_4 + \frac{1}{|\mathbf{p}|} \sqrt{\frac{E - mc^2}{2E}} \gamma \mathbf{p} \,, \qquad (7.43)$$

where $E = \sqrt{m^2 c^4 + |\mathbf{p}|^2 c^2}$, $|\mathbf{p}|^2 = (p^1)^2 + (p^2)^2 + (p^3)^2$.

The equation $\Psi^{FW} = U_{FW}\Psi$ is written in our variables as

$$\begin{pmatrix} \Psi_+^{FW} \\ \Psi_-^{FW} \end{pmatrix} = \tilde{U}_{FW} \begin{pmatrix} \Psi_+ \\ \Psi_- \end{pmatrix} \qquad (7.44)$$

and $\tilde{U}_{FW}$ is given by

$$\tilde{U}_{FW} = \sqrt{\frac{E + mc^2}{2E}} I_8 + \frac{1}{|\mathbf{p}|} \sqrt{\frac{E - mc^2}{2E}} \begin{pmatrix} 0 & \Lambda \\ -\Lambda & 0 \end{pmatrix} \,; \qquad (7.45)$$

$$\Lambda = \begin{pmatrix} \sigma_0 p^3 & \sigma^0 p^1 + i\sigma^2 p^2 \\ \sigma_0 p^1 - i\sigma^2 p^2 & -\sigma^0 p^3 \end{pmatrix}, \quad \Lambda^T = \Lambda \,. \qquad (7.46)$$

This is an orthogonal transformation.

The Dirac equation is written in the new variables as

$$\hbar \begin{pmatrix} \dot{\Psi}_+^{FW} \\ \dot{\Psi}_-^{FW} \end{pmatrix} = \tilde{U}_{FW} H \tilde{U}_{FW}^T \begin{pmatrix} \Psi_+^{FW} \\ \Psi_-^{FW} \end{pmatrix} = E \begin{pmatrix} Y_1 & 0 \\ 0 & -Y_1 \end{pmatrix} \begin{pmatrix} \Psi_+^{FW} \\ \Psi_-^{FW} \end{pmatrix} \qquad (7.47)$$

Note that this makes use of only *one* hyperkahler structure. In fact, this result can be read directly from the transformed Dirac equation under the Foldy-Wouthuysen transformation:

$$i\hbar \dot{\Psi}_{FW} = E\gamma^0 \Psi_{FW} \qquad (7.48)$$

Using (7.5), we get

$$-i \begin{pmatrix} \sigma_0 & 0 \\ 0 & -\sigma_0 \end{pmatrix} \simeq \begin{pmatrix} Y_1 & 0 \\ 0 & -Y_1 \end{pmatrix} \qquad (7.49)$$

and (7.47) follows.

7.2.7   The Cini-Touschek Transformation

The Cini-Touschek transformation is:

$$U_{CT} = \sqrt{\frac{E + |\mathbf{p}|c}{2E}} I_4 - \frac{1}{|\mathbf{p}|} \sqrt{\frac{E - |\mathbf{p}|c}{2E}} \gamma \mathbf{p} \qquad (7.50)$$

and the wave function is transformed as

$$\begin{pmatrix} \psi_+^{CT} \\ \psi_-^{CT} \end{pmatrix} = \tilde{U}_{CT} \begin{pmatrix} \psi_+ \\ \psi_- \end{pmatrix} \tag{7.51}$$

where

$$\tilde{U}_{CT} = \sqrt{\frac{E + |\mathbf{p}|c}{2E}} I_8 - \frac{1}{|\mathbf{p}|} \sqrt{\frac{E - |\mathbf{p}|c}{2E}} \begin{pmatrix} 0 & \Lambda \\ -\Lambda & 0 \end{pmatrix}. \tag{7.52}$$

and Λ is the same as above (7.46). As in the case of the Foldy-Wouthuysen transformation, this is an orthogonal transformation. The Dirac equation is written in the new variables as

$$\hbar \begin{pmatrix} \dot{\psi}_+^{CT} \\ \dot{\psi}_-^{CT} \end{pmatrix} = \tilde{U}_{CT} H \tilde{U}_{CT}^T \begin{pmatrix} \psi_+^{CT} \\ \psi_-^{CT} \end{pmatrix} = \frac{E}{|\mathbf{p}|} \begin{pmatrix} 0 & -\widehat{\mathbf{K}} \\ -\widehat{\mathbf{K}} & 0 \end{pmatrix} \begin{pmatrix} \psi_+^{CT} \\ \psi_-^{CT} \end{pmatrix}. \tag{7.53}$$

Again, note that we need only *one* hyperkahler structure.

As above, this result can be obtained from the transformed Dirac equation under the Cini-Touschek transformation:

$$i\hbar \dot{\Psi}_{CT} = \frac{E}{|\mathbf{p}|} \gamma^0 \gamma \mathbf{p} \, \Psi_{CT} \tag{7.54}$$

Using (7.5), we get

$$- i\gamma^0 \gamma \mathbf{p} \simeq \begin{pmatrix} 0 & -\widehat{\mathbf{K}} \\ -\widehat{\mathbf{K}} & 0 \end{pmatrix} \tag{7.55}$$

and (7.47) follows.

7.2.8 The Mulligan Transformation

Following a recent paper by Mulligan [122], we will consider yet another transformation. Let N be the unitary matrix

$$N = \frac{1}{\sqrt{2}} \begin{pmatrix} \sigma_0 & \sigma_0 \\ -\sigma_0 & \sigma_0 \end{pmatrix}. \tag{7.56}$$

If we apply N to the standard representation of the γ-matrices, the new set of γ-matrices is

$$\tilde{\gamma}^\mu = N \gamma^\mu N^+. \tag{7.57}$$

More explicitly, we have

$$\tilde{\gamma}^0 = \begin{pmatrix} 0 & -\sigma_0 \\ -\sigma_0 & 0 \end{pmatrix}, \quad \tilde{\gamma} = \begin{pmatrix} 0 & \sigma \\ -\sigma & 0 \end{pmatrix}; \tag{7.58}$$

this is the usual spinor representation up to a global minus sign,

$$\tilde{\gamma}^{\mu} = -\gamma^{\mu}_{sp}. \tag{7.59}$$

Note that

$$N\gamma N^+ = \gamma, \qquad [N, U_{CT}] = 0. \tag{7.60}$$

The Mulligan transformation U_M can be understood as the Cini-Touschek transformation applied to the spinor representation of the γ matrices (the original Cini-Touschek transformation was applied to the standard representation of the γ matrices) because N and the Cini-Touschek transformation commute.

In fact, the Cini-Touschek transformation can be written as

$$U_{CT} = \sqrt{\frac{E + |\mathbf{p}|c}{2E}} I_4 + \frac{1}{|\mathbf{p}|}\sqrt{\frac{E - |\mathbf{p}|c}{2E}}\gamma_{sp}\mathbf{p}; \tag{7.61}$$

the Mulligan transformation is similarly written as

$$U_M = \sqrt{\frac{E + |\mathbf{p}|c}{2E}} N - \frac{1}{|\mathbf{p}|}\sqrt{\frac{E - |\mathbf{p}|c}{2E}}N\gamma\mathbf{p}. \tag{7.62}$$

The wave function is transformed as

$$\begin{pmatrix} \psi^M_+ \\ \psi^M_- \end{pmatrix} = \tilde{U}_M \begin{pmatrix} \psi_+ \\ \psi_- \end{pmatrix} \tag{7.63}$$

and $\tilde{U}_M$ is

$$\tilde{U}_M = \sqrt{\frac{E + |\mathbf{p}|c}{4E}} \begin{pmatrix} I_4 & I_4 \\ -I_4 & I_4 \end{pmatrix} - \frac{1}{|\mathbf{p}|}\sqrt{\frac{E - |\mathbf{p}|c}{4E}} \begin{pmatrix} \Lambda & -\Lambda \\ \Lambda & \Lambda \end{pmatrix}. \tag{7.64}$$

As in previous cases, this is once again an orthogonal transformation; the Dirac equation is written in the new variables as

$$\hbar \begin{pmatrix} \dot{\psi}^M_+ \\ \dot{\psi}^M_- \end{pmatrix} = \tilde{U}_M H \tilde{U}^T_M \begin{pmatrix} \psi^M_+ \\ \psi^M_- \end{pmatrix} = \frac{E}{|\mathbf{p}|} \begin{pmatrix} -\widehat{\mathbf{K}} & 0 \\ 0 & \widehat{\mathbf{K}} \end{pmatrix} \begin{pmatrix} \psi^M_+ \\ \psi^M_- \end{pmatrix}. \tag{7.65}$$

We need only *one* hyperkahler structure.

Needless to say, the Dirac equation under Mulligan transformation in quaternionic formulation, can also be obtained in a direct way using (7.5):

$$i\hbar\dot{\psi}_M = \frac{E}{|\mathbf{p}|} \begin{pmatrix} \sigma\mathbf{p} & 0 \\ 0 & -\sigma\mathbf{p} \end{pmatrix} \psi_M, \quad -i \begin{pmatrix} \sigma\mathbf{p} & 0 \\ 0 & -\sigma\mathbf{p} \end{pmatrix} \simeq \begin{pmatrix} -\widehat{\mathbf{K}} \\ \widehat{\mathbf{K}} \end{pmatrix} \tag{7.66}$$

7.3 The Taub-NUT Metric

The Taub-NUT four-dimensional space-time can be obtained from Euclidean eight-dimensional one through a momentum map construction; the HKLR theorem [92] guarantees the hyperkahler structure of $\mathbf{R}^8$ descends to a hyperkahler structure in the Taub-NUT space. Here we present a detailed and fully explicit construction of the hyperkahler structure of a space-time with a Taub-NUT metric.

A specially interesting example of non-trivial hyperkahler manifold is provided by Taub-NUT (Newman, Unti, and Tamburino) space-time [118, 119, 126, 141]; this is physically relevant, and of the minimal dimension (four) for hyperkahler manifolds. It provides an explicit example of nontrivial hyperkahler manifold, which can also be used as a test case in the study of hyperhamiltonian dynamics [60] outside of the standard Euclidean cases.

It appears that the HKLR theorem [92] is not accompanied, in the literature, by many examples for which the hyperkahler structure in the reduced manifold are explicitly provided. We will provide in this section an explicit expression through a direct computation. For more details see [67].

7.3.1 Construction of the Taub-NUT Metric

We will shortly discuss in this section the construction of the Tab-NUT metric. Let E be a complex line bundle, with fiber $\mathbf{C}$, and base the interval $[0, \ell]$. The coordinates in the bundle are (z, x), where $z \in \mathbf{C}$ and $x \in [0, \ell]$.

We define a connection:

$$\frac{\mathrm{d}}{\mathrm{d}x} - it_0 \tag{7.67}$$

where t_0 is a Hermitian endomorphism of $\mathbf{C}$, depending on $x \in [0, \ell]$, (in fact, a real number for each $x \in [0, \ell]$) and three more Hermitian endomorphisms of $\mathbf{C}$, $(t_1(x), t_2(x), t_3(x))$. Finally, let us consider two linear maps from the fiber at $x = 0$ to the fiber at $x = \ell$, (b_0), and viceversa (b_ℓ) (all this information can be encoded in a bow diagram [39]).

We consider the action of a local $U(1)$ as a gauge group, in the following way (we skip the details since they are very well known): if $g(x) \in U(1)$

$$t_0(x) \rightarrow g^{-1}t_0 g + ig^{-1}\frac{\mathrm{d}g(x)}{\mathrm{d}x}$$
$$t_i(x) \rightarrow g^{-1}t_i g \ (i = 1, 2, 3)$$
$$b_0 \rightarrow g^{-1}(0)b_0 g(\ell)$$
$$b_\ell \rightarrow g^{-1}(\ell)b_\ell g(0) \tag{7.68}$$

where (we consider a nontrivial action at the end points and a linear interpolating function for the interior of the interval):

$$g(x) = e^{if(x)} \in U(1), \quad f(x) = \frac{1}{\ell}((\ell - x)\phi_0 + x\phi_\ell), \quad g(0) = e^{i\phi_0}, \quad g(\ell) = e^{i\phi_\ell}$$
(7.69)

Since the linear maps b_0 and b_ℓ are complex, we will write:

$$b_0 = q_0 + iq_1, \quad b_\ell = q_2 + iq_3$$
(7.70)

and if $\theta = \phi_\ell - \phi_0$, we get the following action (all the coordinates are real):

$$t_0 \to t_0 - \frac{\theta}{\ell}, \quad t_i \to t_i, \quad i = 1, 2, 3$$
$$q_0 \to q_0 \cos\theta - q_1 \sin\theta$$
$$q_1 \to q_0 \sin\theta + q_1 \cos\theta$$
$$q_2 \to q_2 \cos\theta + q_3 \sin\theta$$
$$q_3 \to -q_2 \sin\theta + q_3 \cos\theta$$
(7.71)

Using the momentum map associated to this action we can consider the coordinates $t_i, i = 0, 1, 2, 3$ as constants (that is, not depending on x). The Euclidean metric is:

$$\begin{aligned}ds^2 &= \int_0^\ell (dt_0^2 + dt_1^2 + dt_2^2 + dt_3^2)dx + dq_0^2 + dq_1^2 + dq_2^2 + dq_3^2 \\ &= \ell(dt_0^2 + dt_1^2 + dt_2^2 + dt_3^2) + dq_0^2 + dq_1^2 + dq_2^2 + dq_3^2.\end{aligned}$$
(7.72)

We can also consider this space as the sum of two copies of $\mathbf{R}^4$. In the second copy, with coordinates q_i, we introduce quaternionic coordinates, and change the variables, first to a polar form (with angle $\psi/2$) and second to the coordinates $r_i, i = 1, 2, 3$ and ψ given by

$$q_0 = -\sqrt{\frac{r + r_1}{2}} \sin\frac{\psi}{2}$$

$$q_1 = \sqrt{\frac{r + r_1}{2}} \cos\frac{\psi}{2}$$

$$q_2 = \frac{1}{\sqrt{2(r + r_1)}}\left(r_2 \cos\frac{\psi}{2} + r_3 \sin\frac{\psi}{2}\right)$$

$$q_3 = \frac{1}{\sqrt{2(r + r_1)}}\left(-r_2 \sin\frac{\psi}{2} + r_3 \cos\frac{\psi}{2}\right)$$
(7.73)

where

$$r = \sqrt{r_1^2 + r_2^2 + r_3^2}.$$
(7.74)

It is a simple task to write the metric in these coordinates.

$$ds^2 = \ell(dt_0^2 + dt^2) + \tfrac{1}{4}\left[\tfrac{1}{r}dr^2 + r(d\psi + \boldsymbol{\sigma}\cdot dr)^2\right] \qquad (7.75)$$

where

$$\mathbf{t} = (t_1, t_2, t_3), \quad \mathbf{r} = (r_1, r_2, r_3), \quad \boldsymbol{\sigma} = (\sigma_1, \sigma_2, \sigma_3) = \left(0, \frac{r_3}{r(r+r_1)}, -\frac{r_2}{r(r+r_1)}\right). \qquad (7.76)$$

Since we will use them in the forthcoming discussions, we will write explicitly the matrix of the metric (in the second copy of $\mathbf{R}^4$) in the coordinates (r_1, r_2, r_3, ψ):

$$G^{(1)} = \frac{r}{4}\begin{pmatrix} \frac{1}{r^2} & 0 & 0 & 0 \\ 0 & \frac{1}{r^2} + \sigma_2^2 & \sigma_2\sigma_3 & \sigma_2 \\ 0 & \sigma_2\sigma_3 & \frac{1}{r^2} + \sigma_3^2 & \sigma_3 \\ 0 & \sigma_2 & \sigma_3 & 1 \end{pmatrix} \qquad (7.77)$$

and the Jacobian matrix $(\partial q/\partial(\mathbf{r}, \psi))$ of the change of coordinates:

$$\Lambda = \frac{1}{2}\sqrt{\frac{r+r_1}{2}}\left(\Lambda_1 \cos\frac{\psi}{2} + \Lambda_2 \sin\frac{\psi}{2}\right) \qquad (7.78)$$

$$\Lambda_1 = \begin{pmatrix} 0 & 0 & 0 & -1 \\ \frac{1}{r} & -\sigma_3 & \sigma_2 & 0 \\ \sigma_3 & \frac{1}{r} + r\sigma_2^2 & r\sigma_2\sigma_3 & r\sigma_2 \\ -\sigma_2 & r\sigma_2\sigma_3 & \frac{1}{r} + r\sigma_3^2 & r\sigma_3 \end{pmatrix} \qquad (7.79)$$

$$\Lambda_2 = \begin{pmatrix} -\frac{1}{r} & \sigma_3 & -\sigma_2 & 0 \\ 0 & 0 & 0 & -1 \\ -\sigma_2 & r\sigma_2\sigma_3 & \frac{1}{r} + r\sigma_3^2 & r\sigma_3 \\ -\sigma_3 & -\frac{1}{r} - r\sigma_2^2 & -r\sigma_2\sigma_3 & -r\sigma_2 \end{pmatrix} \qquad (7.80)$$

The relation between the matrices $G^{(1)}$ and Λ is the usual one (since the matrix of the metric in the cartesian coordinates for the Euclidean space is the identity):

$$G^{(1)} = \Lambda^T \Lambda. \qquad (7.81)$$

We pass to a quotient space where the Taub-NUT metric is the reduction of the Euclidean metric described in the above paragraphs, using the momentum map (associated to the action of the group $U(1)$). The inverse image of 0 under this map is a submanifold of $\mathbf{R}^8$ given by

$$\mathbf{t} = -\frac{1}{2}\mathbf{r} \qquad (7.82)$$

and the metric, with coordinates $(t_0, r_1, r_2, r_3, \psi)$ is

$$ds^2 = \ell dt_0^2 + \frac{1}{4}\left[\left(\frac{1}{r} + \ell\right)d\mathbf{r}^2 + r(d\psi + \boldsymbol{\sigma} \cdot d\mathbf{r})^2\right].$$ (7.83)

The action of the gauge group (7.71) on this manifold is:

$$t_0 \rightarrow t_0 - \frac{\theta}{\ell}$$
$$r_i \rightarrow r_i, \quad i = 1, 2, 3$$
$$\psi \rightarrow \psi + 2\theta$$ (7.84)

with an invariant given by

$$\tau = 2\ell t_0 + \psi, \quad d\psi = d\tau - 2\ell dt_0$$ (7.85)

which yields the following expression for the metric (in the coordinates $(t_0, r_1, r_2, r_3, \tau)$):

$$ds^2 = \ell dt_0^2 + \frac{1}{4}\left[\left(\frac{1}{r} + \ell\right)d\mathbf{r}^2 + r(d\tau - 2\ell dt_0 + \boldsymbol{\sigma} \cdot d\mathbf{r})^2\right]$$ (7.86)

Finally, to remove t_0 (which is not invariant under the group action) we take:

$$dt_0 = \frac{r}{2(1 + \ell r)}(d\tau + \boldsymbol{\sigma} \cdot d\mathbf{r})$$ (7.87)

and we get the Taub-NUT metric

$$ds^2 = \frac{1}{4}\left[\left(\frac{1}{r} + \ell\right)d\mathbf{r}^2 + \frac{1}{\frac{1}{r} + \ell}(d\tau + \boldsymbol{\sigma} \cdot d\mathbf{r})^2\right].$$ (7.88)

We will write the matrix associated to the Taub-NUT metric (in the coordinates (r_1, r_2, r_3, τ)) which will be used in the following subsections:

$$G^{\text{TNUT}} = \frac{1}{4(\ell + \frac{1}{r})}\begin{pmatrix} (\ell + \frac{1}{r})^2 & 0 & 0 & 0 \\ 0 & (\ell + \frac{1}{r})^2 + \sigma_2^2 & \sigma_2\sigma_3 & \sigma_2 \\ 0 & \sigma_2\sigma_3 & (\ell + \frac{1}{r})^2 + \sigma_3^2 & \sigma_3 \\ 0 & \sigma_2 & \sigma_3 & 1 \end{pmatrix}$$ (7.89)

7.3.2 Quotient Hyperkahler Structures

We will discuss in the following sections how to construct a hyperkahler structure in a 4-dimensional manifold with a Taub-NUT metric. This is an example of the construction of hyperkahler spaces as quotients.

Our goal is to construct explicitly the complex structures and the symplectic forms associated to the hyperkahler structure, when the metric is the Taub-NUT metric written in the coordinates we used in Sect. 7.3.1. As in that approach, our starting point will be a standard hyperkhaler structure in $\mathbf{R}^8$. In the following we will refer to standard hyperkahler structures in $\mathbf{R}^{4n}$ which are obtained from standard structures in $\mathbf{R}^4$ (endowed with an Euclidean metric). As we have explained in Sect. 5.4, there are two such standard structures (1.23) and (1.25), differing for their orientation, with corresponding symplectic structures (1.24) and (1.26) respectively, which in matrix form read as:

$$K_\alpha^{(0)} = Y_\alpha, \quad \alpha = 1, 2, 3, \tag{7.90}$$

since the matrix of the Euclidean metric is the identity (in the cartesian coordinates we are using). Note that, in the general case, if G is the matrix of the metric, J_α the matrices of the hyperkahler structures and K_α the matrices of the symplectic forms, the following relations hold:

$$J_\alpha = G^{-1} K_\alpha, \quad \alpha = 1, 2, 3. \tag{7.91}$$

When we change the coordinate system, as we did in Sect. 7.3.1 passing from the Cartesian coordinates (q_0, q_1, q_2, q_3) to the coordinates (r_1, r_2, r_3, ψ), the corresponding, matrices of the metric, symplectic forms and hyperkahler structure change, and we get

$$G^{(1)} = \Lambda^T I_4 \Lambda = \Lambda^T \Lambda, \quad K_\alpha^{(1)} = \Lambda^T K_\alpha^{(0)} \Lambda = \Lambda^T Y_\alpha \Lambda, \quad J_\alpha^{(1)} = \Lambda^{-1} Y_\alpha \Lambda; \tag{7.92}$$

and the general relation still holds

$$J_\alpha^{(1)} = (G^{(1)})^{-1} K_\alpha^{(1)}, \quad \alpha = 1, 2, 3. \tag{7.93}$$

The symplectic form matrices $K_\alpha^{(1)}$ are

$$K_1^{(1)} = \frac{1}{4} \begin{pmatrix} 0 & \sigma_2 & \sigma_3 & 1 \\ -\sigma_2 & 0 & \frac{1}{r} & 0 \\ -\sigma_3 & -\frac{1}{r} & 0 & 0 \\ -1 & 0 & 0 & 0 \end{pmatrix}$$

$$K_2^{(1)} = \frac{1}{4} \begin{pmatrix} 0 & \frac{1}{r} & 0 & 0 \\ -\frac{1}{r} & 0 & -\sigma_2 & 0 \\ 0 & \sigma_2 & 0 & 1 \\ 0 & 0 & -1 & 0 \end{pmatrix} \tag{7.94}$$

$$K_3^{(1)} = \frac{1}{4} \begin{pmatrix} 0 & 0 & -\frac{1}{r} & 0 \\ 0 & 0 & \sigma_3 & 1 \\ \frac{1}{r} & -\sigma_3 & 0 & 0 \\ 0 & -1 & 0 & 0 \end{pmatrix}$$

and the hyperkahler matrices $J_\alpha^{(1)}$,

$$J_1^{(1)} = \begin{pmatrix} 0 & r\sigma_2 & r\sigma_3 & r \\ 0 & 0 & 1 & 0 \\ 0 & -1 & 0 & 0 \\ -\frac{1}{r} & \sigma_3 & -\sigma_2 & 0 \end{pmatrix}$$

$$J_2^{(1)} = \begin{pmatrix} 0 & 1 & 0 & 0 \\ -1 & 0 & 0 & 0 \\ 0 & r\sigma_2 & r\sigma_3 & r \\ \sigma_2 & -r\sigma_2\sigma_3 & -\frac{1}{r} - r\sigma_3^2 & -r\sigma_3 \end{pmatrix} \qquad (7.95)$$

$$J_3^{(1)} = \begin{pmatrix} 0 & 0 & -1 & 0 \\ 0 & r\sigma_2 & r\sigma_3 & r \\ 1 & 0 & 0 & 0 \\ -\sigma_3 & -\frac{1}{r} - r\sigma_2^2 & -r\sigma_2\sigma_3 & -r\sigma_2 \end{pmatrix}$$

and they satisfy the quaternionic relations (1.15).

Since we have two copies of $\mathbf{R}^4$ in our original space $\mathbf{R}^8$ we could consider several combinations of positive or negative oriented hyperkahler structures. However, the procedure will be essentially the same and we will restrict to the case of two positive oriented hyperkahler structures. Hence, in the original $\mathbf{R}^8$ and cartesian coordinates, these matrices are (I_4 is the identity matrix in four dimensions):

$$\mathfrak{G}^{(0)} = \begin{pmatrix} \ell I_4 & 0 \\ 0 & I_4 \end{pmatrix}, \quad \mathfrak{K}_\alpha^{(0)} = \begin{pmatrix} \ell Y_\alpha & 0 \\ 0 & Y_\alpha \end{pmatrix},$$

$$\mathfrak{J}_\alpha^{(0)} = (\mathfrak{G}^{(0)})^{-1} \mathfrak{K}_\alpha^{(0)} = \begin{pmatrix} Y_\alpha & 0 \\ 0 & Y_\alpha \end{pmatrix}, \quad \alpha = 1, 2, 3 \qquad (7.96)$$

After changing the coordinates in the second copy of $\mathbf{R}^4$ we get the following set of matrices:

$$\mathfrak{G}^{(1)} = \begin{pmatrix} \ell I_4 & 0 \\ 0 & G^{(1)} \end{pmatrix}, \quad \mathfrak{K}_\alpha^{(1)} = \begin{pmatrix} \ell Y_\alpha & 0 \\ 0 & K_\alpha^{(1)} \end{pmatrix},$$

$$\mathfrak{J}_\alpha^{(1)} = (\mathfrak{G}^{(1)})^{-1} \mathfrak{K}_\alpha^{(1)} = \begin{pmatrix} Y_\alpha & 0 \\ 0 & J_\alpha^{(1)} \end{pmatrix}, \quad \alpha = 1, 2, 3 \qquad (7.97)$$

7.3.3 Hyperkahler Structure and the Taub-NUT Metric

In the construction of the Taub-NUT metric we reduce an eight dimensional manifold to a four dimensional one. Our aim is to study the reduction of the hyperkahler structure. A direct approach to this problem is to write the metric and the symplectic forms in the new coordinates. We have solved the problem with the metric, but not with the symplectic forms and we do not have an explicit form for the quotient under the action of the gauge group. But we know explicitly the relation between the forms which provides the quotient space (see Eqs. (7.82), (7.85) and (7.87))

$$\mathbf{dt} = -\frac{1}{2}\mathbf{dr}, \quad d\psi = d\tau - 2\ell dt_0, \quad dt_0 = \frac{r}{2(1 + \ell r)}(d\tau + \boldsymbol{\sigma} \cdot \mathbf{dr}) \tag{7.98}$$

and that is the only fact we need to construct the symplectic forms. In an explicit form

$$
\begin{aligned}
dt_0 &= \frac{r}{2(1 + \ell r)}d\tau + \frac{r}{2(1 + \ell r)}\boldsymbol{\sigma} \cdot \mathbf{dr} \\
dt_\alpha &= -\frac{1}{2}dr_\alpha, \quad \alpha = 1, 2, 3 \\
d\psi &= \frac{1}{1 + \ell r}d\tau - \frac{\ell r}{1 + \ell r}\boldsymbol{\sigma} \cdot \mathbf{dr}
\end{aligned}
\tag{7.99}
$$

Before the reduction, the symplectic forms are (corresponding to the matrices $\mathfrak{K}_\alpha^{(1)}$, see (7.97)):

$$
\begin{aligned}
w_1 &= \ell dt_0 \wedge dt_1 + \ell dt_2 \wedge dt_3 + \frac{1}{4}\sigma_2\, dr_1 \wedge dr_2 + \frac{1}{4}\sigma_3\, dr_1 \wedge dr_3 \\
&\quad + \frac{1}{4r}dr_2 \wedge dr_3 + \frac{1}{4}dr_1 \wedge d\psi
\end{aligned}
\tag{7.100}
$$

$$w_2 = \ell dt_0 \wedge dt_3 + \ell dt_1 \wedge dt_2 + \frac{1}{4r}dr_1 \wedge dr_2 - \frac{1}{4}\sigma_2\, dr_2 \wedge dr_3 + \frac{1}{4}dr_3 \wedge d\psi$$

$$w_3 = \ell dt_0 \wedge dt_2 + \ell dt_3 \wedge dt_1 - \frac{1}{4r}dr_1 \wedge dr_3 + \frac{1}{4}\sigma_3\, dr_2 \wedge dr_3 + \frac{1}{4}dr_2 \wedge d\psi$$

and in the quotient space, after substituting (7.99)

$$
\begin{aligned}
w_1 &= \frac{1}{4}dr_1 \wedge d\tau + \frac{1}{4}\sigma_2\, dr_1 \wedge dr_2 + \frac{1}{4}\sigma_3\, dr_1 \wedge dr_3 + \frac{1}{4}\left(\ell + \frac{1}{r}\right)dr_2 \wedge dr_3 \\
w_2 &= \frac{1}{4}dr_3 \wedge d\tau - \frac{1}{4}\sigma_2\, dr_2 \wedge dr_3 + \frac{1}{4}\left(\ell + \frac{1}{r}\right)dr_1 \wedge dr_2 \\
w_3 &= \frac{1}{4}dr_2 \wedge d\tau + \frac{1}{4}\sigma_3\, dr_2 \wedge dr_3 - \frac{1}{4}\left(\ell + \frac{1}{r}\right)dr_1 \wedge dr_3
\end{aligned}
\tag{7.101}
$$

with matrices

$$
K_1^{\text{TNUT}} = \frac{1}{4}
\begin{pmatrix}
0 & \sigma_2 & \sigma_3 & 1 \\
-\sigma_2 & 0 & \ell + \frac{1}{r} & 0 \\
-\sigma_3 & -\left(\ell + \frac{1}{r}\right) & 0 & 0 \\
-1 & 0 & 0 & 0
\end{pmatrix},
$$

$$K_2^{\text{TNUT}} = \frac{1}{4} \begin{pmatrix} 0 & \ell + \frac{1}{r} & 0 & 0 \\ -\left(\ell + \frac{1}{r}\right) & 0 & -\sigma_2 & 0 \\ 0 & \sigma_2 & 0 & 1 \\ 0 & 0 & -1 & 0 \end{pmatrix}, \tag{7.102}$$

$$K_3^{\text{TNUT}} = \frac{1}{4} \begin{pmatrix} 0 & 0 & -\left(\ell + \frac{1}{r}\right) & 0 \\ 0 & 0 & \sigma_3 & 1 \\ \ell + \frac{1}{r} & -\sigma_3 & 0 & 0 \\ 0 & -1 & 0 & 0 \end{pmatrix}$$

Since we know the matrix of the metric (7.89), we can compute the matrices of the hyperkahler structure $(Y_\alpha^{\text{TNUT}} = (G^{\text{TNUT}})^{-1} K_\alpha^{\text{TNUT}})$; here, for ease of notation, we set

$$\lambda = \ell + \frac{1}{r} .$$

$$J_1^{\text{TNUT}} = \begin{pmatrix} 0 & \sigma_2/\lambda & \sigma_3/\lambda & 1/\lambda \\ 0 & 0 & 1 & 0 \\ 0 & -1 & 0 & 0 \\ -\lambda & \sigma_3 & -\sigma_2 & 0 \end{pmatrix} \tag{7.103}$$

$$J_2^{\text{TNUT}} = \begin{pmatrix} 0 & 1 & 0 & 0 \\ -1 & 0 & 0 & 0 \\ 0 & \sigma_2/\lambda & \sigma_3/\lambda & 1/\lambda \\ \sigma_2 & -\sigma_2\sigma_3/\lambda & -\lambda - \sigma_3^2/\lambda & -\sigma_3/\lambda \end{pmatrix}$$

$$J_3^{\text{TNUT}} = \begin{pmatrix} 0 & 0 & -1 & 0 \\ 0 & \sigma_2/\lambda & \sigma_3/\lambda & 1/\lambda \\ 1 & 0 & 0 & 0 \\ -\sigma_3 & -\lambda - \sigma_2^2/\lambda & -\sigma_2\sigma_3/\lambda & -\sigma_2/\lambda \end{pmatrix}$$

which satisfy the quaternionic relations (1.15). It can also be checked that they are covariantly constant under the Levi-Civita connection associated to the Taub-NUT metric.

We have thus explicitly computed the hyperkahler structures, thus implementing the abstract HKLR theorem [92] in the concrete case of Taub-NUT. Note that all these structures reproduce the flat case when $r \to 0$.

Had we chosen as starting point a standard hyperkahler structure in $\mathbf{R}^8 = \mathbf{R}^4 \oplus \mathbf{R}^4$ with different orientations, we would have obtained similar results. In fact, working with a negative orientation we can easily choose a different set of coordinates (7.73) and obtain the same expressions for the hyperkahler and $\mathbf{Q}$-structures.

Appendix A
Holonomy

The holonomy of a manifold [5, 21, 25, 56, 83, 101, 104, 107, 112, 123, 127, 128, 133–135] plays a fundamental role in Differential Geometry and its applications in Physics. Its connection to curvature converts this concept into a useful tool in the description of many physical systems.

Holonomy is also a fundamental ingredient in the study of hyperkahler manifolds, and it is thus worth giving a discussion of it.

In the sequel we will shortly introduce the definitions and main properties in what concerns its use in the description of the manifolds to be considered in this book. We will closely follow the exposition in [106] for the general concepts on connections and [100, 104, 133] for those questions related to holonomy of Riemannian manifolds and Berger's list.

A.1   Connections

Let M be a differentiable manifold and $P(M, G)$ a principal fiber bundle (G being the structure Lie group).

A *connection* Γ [106] selects, for each point $u \in P$, a horizontal space; that is, a subspace such that the whole tangent space at u is the direct sum of the vertical space (tangent to the fiber, G) and such horizontal space. The horizontal subspace at another point (which is obtained as the right action of G over M) is given by the action of the corresponding tangent map of this right action (covariance). This choice should be differentiable. Any paracompact fiber bundle admits a connection.

The connection can be implemented as follows. Given a vector field X in the tangent space at $u \in P$, it has a unique associated vertical vector field (chosen by the connection). Since the vertical vector fields are the fundamental vector fields corresponding to the Lie algebra of G, we define a one-form in P with values in the Lie algebra, such that $\omega(X)$ is this vector field in the Lie algebra. The form ω is called the **connection form** of the connection Γ.

© Springer International Publishing AG 2017
G. Gaeta and M.A. Rodríguez, *Lectures on Hyperhamiltonian
Dynamics and Physical Applications*, Mathematical Physics Studies,
DOI 10.1007/978-3-319-54358-1

A connection allows to define the horizontal lift of any vector field in P. The horizontal lift is invariant under the action of the right action.

The introduction of a connection in a manifold gives rise to the notion of parallelism and parallel transport, comparing vector fields (in fact vertical fibers) at different points of the manifold. If $x(t)$ is a differentiable curve on the manifold M we can define its horizontal lift (corresponding to the horizontal lift of its tangent vector field) passing through point of the vertical fiber. We can assign to any point of the fiber through $x(t_0)$ the corresponding point (along the lifted curve) in the fiber through $x(t_1)$. This correspondence will be called the *parallel displacement* along the curve.

A.2 Holonomy Groups

Given a connection Γ in a principal fiber bundle, we consider the set of closed curves (loops) starting and ending at a point $x \in M$. We can compose two such curves (considering the curve formed by one of them and then the other) at a given point. For each curve we consider the parallel displacement along it, which provides an isomorphism of the fiber $\pi^{-1}(x)$ onto itself. Considering the composition of curves, these isomorphisms form a group, which is (isomorphic to) a subgroup of the group G, the *holonomy group* $\Phi(u)$ of the connection with reference point x. If we restrict the loops to those which are homotopic to zero, we get a subgroup of the holonomy group, the *restricted holonomy group* (again, with reference point x).

If two points in the manifold M can be joined by a curve, the corresponding holonomy (and restricted) groups are isomorphic. In particular, if the manifold is connected, the holonomy groups at each point are conjugate.

It can be shown that the restricted holonomy group is a connected Lie subgroup of the structure group G, and a normal subgroup of the holonomy group at a given reference point.

A.3 Curvature

There is a remarkable relation between the curvature of a manifold and the holonomy group. Let ρ be a representation of the group G on the endomorphisms of a finite dimensional vector space. A *pseudotensorial form* φ of degree r of type (ρ, V) on a bundle P, is a V-valued r-form satisfying $R^*\varphi = \rho(a^{-1})\varphi$, where $a \in G$. If the form is zero when applied to a horizontal vector field, it is called *tensorial*.

If φ is a pseudotensorial form of degree r, the form φ_h, defined over the horizontal part of vector fields, is tensorial. It turns out that $d\varphi$ is also a pseudotensorial form (of degree $r + 1$) and $D\varphi = (d\varphi)_h$ is a tensorial form of degree $r + 1$. This form is called the *exterior covariant derivative* of the form φ.

If we consider the adjoint representation of the group G on its Lie algebra, the connection form ω is a tensorial one-form (of type ad). The *curvature* of a connection is the exterior covariant derivative $\Omega = D\omega$ of the tensorial form ω. The curvature satisfies the structure equation

$$d\omega = -\frac{1}{2} [\omega, \omega] + \Omega \tag{A.1}$$

and the Bianchi identity

$$D\Omega = 0. \tag{A.2}$$

The close relation between the structure group G and the holonomy group is given by the

Theorem (Ambrose-Singer) *Let $P(M, G)$ be a principal fiber bundle with a connected and paracompact base manifold. If Γ is a connection in P, given $u \in P$, the set of points which can be joined to u through a horizontal curve in P, $P(u)$, is a reduced bundle with structure group equal to the holonomy group of the connection at u. The connection in P can be reduced to a connection in $P(u)$.*

Definition A.6 The bundle $(P(u), \Phi(u))$ is called the *holonomy bundle* through u.

The Ambrose-Singer theorem essentially states that the Lie algebra of the holonomy group is a subalgebra of the Lie algebra of the structure group. More precisely, if Γ is a connection in $P(M, G)$, with curvature form Ω, then the Lie algebra of the holonomy group at a given point $u \in P$ is equal to a subspace of the Lie algebra of G, spanned by elements $\Omega_v(X, Y)$, with v an element in the reduced bundle $P(u)$ and X, Y horizontal vectors at v.

This implies that any connected Lie group can be realized as the holonomy group of a connection in a trivial bundle.

In a trivial principal fiber bundle $P \times M$ we can construct a (canonical) flat connection, defining as horizontal space the tangent space to the subbundle $M \times \{a\}$ in each point $a \in G$. The connection form is given by projecting the canonical one-form of G and has zero curvature. In this sense, a connection in a principal fiber bundle is flat if it is given locally by a canonical flat connection. It can be shown that this condition is equivalent to the zero curvature condition.

A.4 Connections in Vector Bundles

Consider a principal vector bundle $P(M, G)$ and $E(M, V, G, P)$, the associated vector bundle, where V is a vector space of finite dimension and G acts on V through a representation ρ of endomorphisms in V. If Γ is a connection in P, φ a section $\varphi : M \to E$, and x_t a curve in M, we can define the *covariant derivative* $\nabla_{\dot{x}_t}$ of φ along the curve x_t, using the parallel displacement given by the connection. The covariant derivative is zero if the curve $\varphi(x_t)$ is horizontal, or in the other words, the section is parallel.

In the bundle of linear frames P over M, we can define in the usual way the canonical $\mathbf{R}^n$-valued one-form θ, which is a tensorial one-form of type $(GL(n, \mathbf{R}), \mathbf{R}^n)$. A *linear connection* is a connection in the bundle of linear frames of P over M. Then we can define $\Theta = D\theta$, which is called the *torsion* of the connection. The canonical one-form, the torsion, and the curvature satisfy the structure equations of the connection and the Bianchi identity. The torsion (T) and curvature (R) tensor fields can be defined in terms of the covariant derivative in the usual way,

$$T(X, Y) = \nabla_X Y - \nabla_Y X - [X, Y] \,, \tag{A.3}$$

$$R(X, Y) Z = [\nabla_X, \nabla_Y] Z - \nabla_{[X,Y]} Z \,. \tag{A.4}$$

The linear connections can be extended to affine connections, considering the vector space $\mathbf{R}^n$ as an affine space.

A.5   Riemannian Holonomy Groups

If (M, g) is a manifold endowed with a Riemannian metric g, the holonomy group of (M, g) is the holonomy group for the Levi-Civita connection associated to g. Since the metric is invariant under the connection (the covariant derivative of g is zero), the holonomy group is a subgroup of the invariance group of the metric, which is (isomorphic to) an orthogonal group. It is independent of the reference point.

For a given holonomy group, the curvature has to satisfy some constraints, apart from those coming from the Bianchi identity. This is the basis of Berger's classification of Riemannian holonomy groups. In his classification, Berger excludes the reducible Riemannian manifolds (which can be studied in terms of the irreducible ones) and those which are symmetric spaces (which are completely determined as homogeneous spaces of Lie groups) and considers simply-connected manifolds.

In fact these characteristics impose severe constraints on the holonomy group. Since the space is simply-connected, the holonomy group (which, we recall, is a Lie group) is connected. The action of the holonomy group (on $\mathbf{R}^n$, where n is the dimension of the manifold) is irreducible. And, finally, since we have excluded the symmetric spaces, the covariant derivative of the curvature tensor is non null.

With these restrictions, Berger provides the following list.

First of all, the holonomy group can be equal to the orthogonal group (the invariance group of the metric). This is the case related to real numbers (where so(n) acts as automorphisms).

Apart from this case, there are six more "special holonomies", when the holonomy group is a proper subgroup of the orthogonal group. Two of them correspond to $n = 7$ when the holonomy group is G_2, the lowest dimension exceptional Lie group, a proper subgroup of SO(8) (in terms of Lie algebras, this is the case $G_2 \subset B_3$, an exceptional case appearing in Dynkin's classification of maximal semisimple Lie algebras, which is also in the basis of the Berger's classification). There is another isolated case, in dimension 8, the group $Spin(7) \subset SO(8)$ (these two groups are related to automorphism groups of octonions).

The other cases correspond to even dimension ($n = 2m$ or $n = 4m$), and to complex or quaternions cases. If $n = 2m \geq 4$, the holonomy group can be $U(m)$ or $SU(m)$ (automorphism groups of complex numbers). If $n = 4m \geq 8$ the holonomy group can be $Sp(m) \subset SO(4m)$ or $\times sp(1) Sp(m) \times Sp(1) \subset SO(4m)$ (automorphism groups of quaternions).[1]

Although Berger did not provide an example of a manifold for each of these groups, these manifolds have been constructed along the years, and for any of the groups in Berger list we know a manifold which has this as holonomy group.

Finally, we recall the main types of manifolds as characterized by their holonomy:

1. A (generic) *Kahler* manifold, that is a complex manifold with a Kahler metric, has as holonomy group $U(m)$.
2. *Calabi-Yau* manifolds have as holonomy group $SU(m)$. They are Kahler manifolds which locally (or, if they are simply-connected, globally) can be considered as Kahler manifolds which are Ricci-flat.
3. *Hyperkahler* manifolds have as holonomy group the direct $Sp(m) \times Sp(1)$. The corresponding metric is Kahler and Ricci-flat.
4. *Quaternionic* manifolds have holonomy group $Sp(2m)$. They are not Kahler manifolds – in fact, they are not Ricci flat – although they are Einstein manifolds (the Ricci tensor is proportional to the metric).

---

[1] Note that we are defining $Sp(m)$ as the group of unitary symplectic matrices in dimension $2m$.

Appendix B
Variational Principles

It is well known that Hamiltonian dynamics admits a variational formulation [10, 12, 109, 111] (note this is formulated in the extended phase manifold). The same holds for hyperhamiltonian dynamics, albeit in this case the underlying variational principle is of a less standard type, based not on one-forms but on forms of the highest possible (not leading to trivial behavior) order; that is order $N - 1$ for a manifold M of dimension N and hence extended phase manifold $M \times \mathbf{R}$ of dimension $N + 1$. This involves Hodge duality in its formulation.

We provide here a brief description of the variational formulation of hyperhamiltonian dynamics, based on such a "maximal order variational principle"; a more detailed discussion of these is provided in [62–64].

B.1 Maximal Order Variational Principles

We consider the *extended phase manifold* $\widetilde{M} = M \times \mathbf{R}$, where the $\mathbf{R}$ factor should be thought as corresponding to the time variable. This is a trivial fiber bundle $\theta : \widetilde{M} \to \mathbf{R}$. On any local chart $U \subseteq M$, we can describe U as a bundle $\pi : U \to B$ over a $(4n - 1)$-dimensional manifold B, and hence $\widetilde{U} := U \times \mathbf{R}$ has the structure of a double fibration

$$U \times \mathbf{R} \xrightarrow{\pi} B \times \mathbf{R} \xrightarrow{\tau} \mathbf{R} ;$$

here of course $\pi \circ \tau = \theta$.

We denote by $\Gamma(\pi)$, $\Gamma(\tau)$, $\Gamma(\theta)$ the set of smooth sections for the bundles π, τ, θ respectively; and by $\mathcal{V}(\pi)$ the set of vertical vector fields for the fibration π (note these cannot have any component along ∂_t). For a vector field $V \in \mathcal{V}(\pi)$, we denote by $\psi_V(.; s)$ the flow generated by V.

Given a section $\sigma \in \Gamma(\pi)$, the *variation* of σ under $V \in \mathcal{V}(\pi)$ is the section $\widetilde{\psi}_V(s) = V(\sigma, s)$. Note that, as V does not have components along ∂_t, we are considering isochronous variations.

© Springer International Publishing AG 2017
G. Gaeta and M.A. Rodríguez, *Lectures on Hyperhamiltonian Dynamics and Physical Applications*, Mathematical Physics Studies,
DOI 10.1007/978-3-319-54358-1

We will now consider again the form ϑ defined above, see (2.13), and a manifold $C \subset B$ with boundary ∂C. We consider the functional $I : \Gamma(\pi) \to \mathbf{R}$ defined (with σ^* the pullback of σ) as

$$I(\sigma) := \int_C \sigma^*(\vartheta) . \tag{B.1}$$

We will restrict our attention to vertical vector fields in

$$\mathcal{V}_C(\pi) := \{V \in \mathcal{V}(\pi) : V|_{\partial C} = 0\} \subset \mathcal{V}(\pi) ;$$

these are just vector fields vanishing on the boundary of C, as customary in the calculus of variations.

We say that a section $\sigma \in \Gamma(\pi)$ is *extremal for I* if and only if

$$\frac{d}{ds} \left[\int_C [\widetilde{\psi}_V(\sigma, s)]^* (\vartheta) \right]_{s=0} = 0 \quad \forall V \in \mathcal{V}_C(\pi) . \tag{B.2}$$

It is a standard result in variational analysis (see e.g. [86]) that

Theorem *A section σ is extremal for I if and only if*

$$\sigma^*(V \lrcorner d\vartheta) = 0 \quad \forall V \in \mathcal{V}_C(\pi) . \tag{B.3}$$

Remark B.1 Note that $\mathcal{V}(\pi)$ is a two-dimensional module over the algebra of smooth functions $f : \widetilde{M} \to \mathbf{R}$; if $\{V_1, V_2\}$ generate this module, the condition (B.3) can be written as

$$\sigma^*(V_1 \lrcorner d\vartheta) = 0 , \quad \sigma^*(V_2 \lrcorner d\vartheta) = 0 . \tag{B.4}$$

The latter is independent of the choice of $C \subset B$. $\odot$

Remark B.2 The Theorem given above can be reformulated in the language of *Cartan ideals* (i.e. ideals of differential forms) [26, 35, 79]. With V_1, V_2 as in the previous Remark 1, $\sigma \in \Gamma(\pi)$ is extremal for I if and only if s is an integral manifold for the ideal $\mathcal{J}$ generated by $\beta_1 = V_1 \lrcorner d\vartheta$ and $\beta_2 = V_2 \lrcorner d\vartheta$. In this sense, the variational principle I (that is, $\delta I = 0$) is associated with the ideal $\mathcal{J}$, and conversely. $\odot$

B.2 Variational Characterization of Hyperhamiltonian Dynamics

We can now state and prove our result concerning the variational characterization of the hyperhamiltonian vector field associated to the Hamiltonians $\mathcal{H}^\alpha$ on a hyperkahler manifold M, or more precisely on the hypersymplectic manifold $(M, g; \omega_1, \omega_2, \omega_3)$.[2]

Lemma *Let X be the hyperhamiltonian vector field generated by the Hamiltonians $\mathcal{H}^\alpha$; for ϑ and I as above, let $\mathcal{J}$ be the ideal associated with I. Then the characteristic distribution $D(\mathcal{J})$ for $\mathcal{J}$ is one dimensional, and it is generated by the vector field $Z = \partial_t + X$.*

Proof In order to prove this result (following [GMhh]), we start by remarking that if η is a (non zero) N-form on the $(N + 1)$ dimensional manifold $\widetilde{M}$, and Y, V_1, V_2 are three independent (and non zero) vector fields on M, then

$$ V_1 \lrcorner (Y \lrcorner \eta) \;=\; 0 \;=\; V_2 \lrcorner (Y \lrcorner \eta) $$

if and only if $Y \lrcorner \eta = 0$. Moreover, the space of vector field Y satisfying this relation is a one-dimensional module over the set of smooth functions $f : \widetilde{M} \to \mathbf{R}$.

In fact, let $\{x^0, ..., x^N\}$ be local coordinates in $\widetilde{M}$, with $\Omega = \mathrm{d}x^0 \wedge ... \wedge \mathrm{d}x^N$; take, with no loss of generality, $X = \partial_0$, $V_1 = \partial_1$, $V_2 = \partial_2$. Then,

$$ \eta \;=\; \sum_{k=0}^{N} c_k \, (\partial_k \lrcorner \Omega) \, . $$

Now the conditions $\partial_j \lrcorner (\partial_0 \lrcorner \eta) = 0$ for $j = 1, 2$ imply, respectively, $c_k = 0$ for $k \neq 0, 1$ and for $k \neq 0, 2$. Thus requiring both conditions yields

$$ \eta \;=\; c_0 \, (\partial_0 \lrcorner \Omega) \;=\; c_0 \, (X \lrcorner \Omega) \, . $$

Needless to say, this satisfies $X \lrcorner \eta = 0$, and conversely $Y \lrcorner \eta = 0$ implies $Y = f X$ with f a smooth function.

The ideal $\mathcal{J}$ generated by $\{\psi_1, \psi_2\}$ with $\psi_i = (V_i \lrcorner \eta)$ is non-singular and admits a one dimensional characteristic distribution $D(\mathcal{J})$: this is precisely the one provided by vector fields X satisfying $X \lrcorner \eta = 0$. In fact, ψ_i are $(N - 1)$-forms, and the condition $(X \lrcorner \psi_i) \in \mathcal{J}$ is equivalent to $X \lrcorner \psi_i = 0$, and we just have a reformulation of the above algebraic remark (note this implies that the space of X such that $X \lrcorner \eta = 0$ has constant dimension, i.e. $\mathcal{J}$ is non singular). Now, it suffices to specialize this discussion to the case where $\eta = \mathrm{d}\vartheta$, and use the "alternative" characterization of X, i.e. the one in terms of Z satisfying (2.14), to conclude that indeed we have a variational characterization of the hyperhamiltonian vector field X. △

---

[2] We recall that $\mathcal{H}^\alpha$ and the ω_α enter in the definition of ϑ, hence of I.

Finally, we note that it follows from this that the vector field Z defined in (2.14) is everywhere tangent to integral manifolds of $\mathcal{J}$, i.e. extremal sections for I. Moreover, as shown in [62, 63] the [$(4n - 1)$-dimensional extremal sections] for I can be described by assigning their value on a suitable $(4n - 2)$-dimensional manifold and "pulling them" along integral curves of Z.

We refer the interested reader to [62–64] for more details.

Appendix C
Quaternions and Quantum Mechanics

Since the invention/discovery of quaternions by Hamilton in 1843, their use in Physics and Mathematics have attracted the interest of many researchers, with periods of intense activity followed by long periods in which this interest has declined (see in [Al86, Al89] some historical remarks of how the theory was received by the scientific community, and in [AJ93] an essay about its role in Physics).

It is impossible to give a list of all the studies and applications that have been achieved in this field, but we would like, with a view at the goal of our work, to point out some of the most recent developments. This in particular for those referring to the application of quaternionic structures in *Quantum Mechanics*, a topic which we did not deal with at all. (We thank a Reviewer for urging us to mention this topic.)

Given the "historical" character of this Appendix, we provide a separate list of references for it.

One of the earliest hints for the use of quaternions in Quantum Mechanics is discussed in the paper by Birkhoff and von Neumann [BN36]. In this paper, it is pointed out that to develop a propositional calculus (providing a rigorous foundation of Quantum Mechanics), one could use any field of numbers with an involutive anti-isomorphism, and these include real, complex and quaternions, although no further development is made on this comment.

In a series of (much later) papers, Finkelstein, Jauch, Schiminovich and Speiser [FJ62] developed a quantum mechanics based on quaternions, where complex quantities were replaced by quaternions. The reasons for this change of the number field are described in their work and can essentially be summarized as the possibility of implement the propositional calculus underlying quantum mechanics using quaternions, as Birkhoff and von Neumann had suggested. In the first paper of the series, the use of quaternions is justified and the difficulties of the anticommutative product are discussed. When they apply this theory to the study of quantum systems, the concept of covariance appears and, therefore, the need to introduce a notion of parallel transport on quaternions and their consequences, covariant derivative and curvature [FJ63a].

© Springer International Publishing AG 2017
G. Gaeta and M.A. Rodríguez, *Lectures on Hyperhamiltonian Dynamics and Physical Applications*, Mathematical Physics Studies,
DOI 10.1007/978-3-319-54358-1

These articles [FJ62, FJ63a, FJ63b] were followed by a great amount of research. See, as a very sketchy list [BG76, BS05, LD04, LD12, HB84, Ho93, Ho96, Ho97, NJ87, NV67, SS97, TM96, TT03] and a rather complete bibliography in [GH08].

We will now briefly mention the works by S. Adler, see [Ad85, Ad86, Ad96, AM97], and in particular [Ad95].

In his work, he carries out a careful analysis of the problems that appear when using quaternions to represent the physical interesting quantities associated to quantum systems, in particular the difficulties derived from the non-commutativity of the quaternion product.

In this sense, it is necessary to choose a multiplication (right or left) for the product by scalars in the corresponding Hilbert space. The spectral theory is not the same as the complex case; for instance, in the study of anti-selfadjoint operators which are basic tools in the description of symmetries. Although the quaternions can be represented by a pair of complex numbers, the quantum mechanics that we can construct with quaternions is not equivalent to a complex mechanics with wave functions having two components. The last question highlighted by Adler is that associativity plays a fundamental role, which excludes the extension of his work to the field of octonions. Adler's book makes a complete journey through quantum theory from its foundations, through non-relativistic Quantum Mechanics to Quantum Field Theory.

References for Appendix C

[Ad85] S.L. Adler, Quaternionic quantum field theory. Phys. Rev. Lett. **55**, 783–786 (1985)

[Ad86] S.L. Adler, Quaternionic quantum field theory. Commun. Math. Phys. **104**, 611–656 (1986)

[Ad95] S.L. Adler, *Quaternionic Quantum Mechanics and Quantum Fields* (Oxford UP, Oxford, 1995)

[Ad96] S.L. Adler, Projective group representations in quaternionic Hilbert space. J. Math. Phys. **37**, 2352–2360 (1996)

[AM97] S.L. Adler, A.C. Millard, Coherent states in quaternionic quantum mechanics. J. Math. Phys. **38**, 2117–2126 (1997)

[Al86] S.L. Altmann, *Rotations, Quaternions and Double Groups* (Clarendon Press, Oxford, 1986)

[Al89] S.L. Altmann, Hamilton, Rodrigues, and the quaternion scandal. Math. Mag. **62**, 291–308 (1989)

[AJ93] R. Anderson, G.C. Joshi, Quaternions and the heuristic role of mathematical structures in Physics. Phys. Essays **6**, 308–319 (1993)

[BG76] N. Backhouse, P. Gard, On the tensor representation for compact groups. J. Math. Phys. **17**, 2098–2100 (1976)

[BN36] G. Birkhoff, J. von Neumann, The logic of quantum mechanics. Ann. Math. **37**, 823–843 (1936)

[BS05] A. Blasi, G. Scolarici, L. Solombrino, Alternative descriptions in quaternionic quantum mechanics. J. Math. Phys. **46**, 042104 (2005)

[FJ62] D. Finkelstein, J.M. Jauch, S. Schiminovich, D. Speiser, Foundations of quaternion quantum mechanics. J. Math. Phys. **3**, 207–220 (1962)

[FJ63a] D. Finkelstein, J.M. Jauch, D. Speiser, Quaternionic representations of compact groups. J. Math. Phys. **4**, 136–140 (1963)

[FJ63b] D. Finkelstein, J.M. Jauch, S. Schiminovich, D. Speiser, Principle of general Q-covariance. J. Math. Phys. **4**, 788–796 (1963)

[GH08] A. Gsponer, J.-P. Hurni, in Quaternions in Mathematical Physics (1): Alphabetical bibliography (2008). arXiv:math-ph/0510059v4

[HB84] L.P. Horwitz, L.C. Biedenharn, Quaternion quantum mechanics: second quantization and gauge fields. Ann. Phys. **157**, 432488 (1984)

[Ho93] L.P. Horwitz, Some spectral properties of anti-self-adjoint operators on a quaternionic Hilbert space. J. Math. Phys. **34**, 3405–3419 (1993)

[Ho96] L.P. Horwitz, Hypercomplex quantum mechanics. Found. Phys. **26**, 851–862 (1996)

[Ho97] L.P. Horwitz, Schwinger algebra for quaternionic quantum mechanics. Found. Phys. **27**, 1011–1034 (1997)

[LD04] S. de Leo, G. Ducati, Quaternionic differential operators. J. Math. Phys. **42**, 2236–2265 (2001)

[LD12] S. de Leo, G. Ducati, Delay time in quaternionic quantum mechanics. J. Math. Phys. **53**, 022102 (2012)

[NJ87] C.G. Nash, G.C. Joshi, Composite systems in quaternionic quantum mechanics. J. Math. Phys. **28**, 2883–2885 (1987)

[NV67] S. Natarajan, K. Viswanath, Quaternionic representations of compact metric groups. J. Math. Phys. **8**, 582–589 (1967)

[SS97] G. Scolarici, L. Solombrino, Quaternionic representations of magnetic groups. J. Math. Phys. **38**, 1147–1160 (1997)

[TM96] T. Tao, A.C. Millard, On the structure of projective group representations in quaternionic Hilbert space. J. Math. Phys. **37**, 5848–5857 (1996)

[TT03] K. Thirulogasanthar, S. T. Ali, "Regular subspaces of a quaternionic Hilbert space from quaternionic Hermite polynomials and associated coherent states". J. Math. Phys. 54, 013506 (2013)

References

1. D.V. Alekseevskii, Compact quaternion spaces. Funct. Anal. Appl. **2**, 97–105 (1968)
2. D.V. Alekseevskii, Riemannian spaces with exceptional holonomy groups. Funct. Anal. Appl. **2**, 106–114 (1968)
3. D.V. Alekseevskii, S. Marchiafava, Quaternionic structures on a manifold and subordinated structures. Ann. Mat. Pura Appl. **171**, 205–273 (1996)
4. D.V. Alekseevskii, S. Marchiafava, M. Pontecorvo, Compatible almost complex structures on quaternion Kähler manifolds. Ann. Glob. Anal. Geom. **16**, 419–444 (1998)
5. W. Ambrose, I.M. Singer, A theorem on holonomy. Trans. A.M.S. 75, 428–443 (1953)
6. I. Anderson, M. Fels, P. Vassiliou, On Darboux integrability, in SPT2007-Symmetry and Perturbation Theory, ed. G. Gaeta, R. Vitolo, S. Walcher (World Scientific, Singapore, 2008)
7. D.V. Anosov, V.I. Arnold, *Dynamical Systems I. Ordinary Differential Equations and Smooth Dynamical Systems*. Encyclopaedia of Mathematical Sciences. vol. 1 (Springer 1988)
8. D. Anselmi, P. Fré, Topological σ-models in four dimensions and triholomorphic maps. Nucl. Phys. B **416**, 255–300 (1994)
9. I. Antoniadis, B. Pioline, Higgs branch, hyperkähler quotients and duality in SUSY N=2 Yang-Mills theories. Int. J. Mod. Phys. A **12**, 4907–4932 (1997)
10. V.I. Arnold, *Mathematical Methods of Classical Mechanics*, 2nd edn. (Springer, Berlin, 1989)
11. V.I. Arnold, *Geometrical methods in the theory of differential equations* (Springer, Berlin, 1983)
12. V.I. Arnold, V.V. Kozlov, A.I. Neishtadt, *Dynamical Systems III. Mathematical Aspects of Classical and Celestial Mechanics*, vol. 3, 3rd edn., Encyclopaedia of Mathematical Sciences (Springer, Berlin, 2006)
13. M.F. Atiyah, Hyper-Kähler manifolds, in *Complex Geometry and Analysis*, ed. by V. Villani. Lecture Notes in Mathematics, vol. 1422 (Springer, Berlin, 1990)
14. M.F. Atiyah, N.J. Hitchin, *The Geometry and Dynamics of Magnetic Monopoles* (Princeton University Press, Princeton, 1988)
15. M. Audin, Intégrabilité et non-intégrabilité de systèmes Hamiltoniens. Séminaire Bourbaki **43**, 113–135 (2000)
16. M. Audin, *Hamiltonian Systems and their Integrability* (S.M.F, Paris, 2008)
17. A. Baker, Matrix Groups: An Introduction to Lie Group Theory (Springer, London, 2002)
18. W. Ballmann, *Lectures on Kähler Manifolds, ESI Lectures in Mathematics and Theoretical Physics* (European Mathematical Society Publishing, Zürich, 2006)
19. V. Bargmann, On a Hilbert space of analytic functions and an associated integral transform. Commun. Pure Appl. Math. **14**, 187–214 (1961)

© Springer International Publishing AG 2017

G. Gaeta and M.A. Rodríguez, *Lectures on Hyperhamiltonian*
Dynamics and Physical Applications, Mathematical Physics Studies,
DOI 10.1007/978-3-319-54358-1

20. G. Benettin, L. Galgani, A. Giorgilli, A proof of Kolmogorov's theorem on invariant tori using canonical transformations defined by the Lie method. Nuovo Cimento B **79**, 201–223 (1984)
21. M. Berger, Sur les groupes d'holonomie homogènes de variétés à connexion affine et des variétés riemanniennes. Bull. Soc. Math. France **83**, 279–330 (1955)
22. J.D. Bjorken, S.D. Drell, *Relativistic Quantum Mechanics* (McGraw-Hill, New York, 1964)
23. H.W. Broer, Bifurcations of singularities in volume preserving vector fields. Ph.D. Thesis, Groningen, 1979
24. H.W. Broer, Formal normal form theorems for vector fields and some consequences for bifurcations in the volume preserving case, in *Dynamical Systems and Turbulence*, vol. 898, Lecture Notes in Mathematics, ed. by D.A. Rand, L.S. Young (Springer, Berlin, 1981)
25. R. Bryant, Recent advances in the theory of holonomy. Séminaire Bourbaki **41**, 351–374 (2000)
26. R.L. Bryant, S.S. Chern, R.B. Gardner, H.L. Goldschmidt, P.A. Griffiths, *Exterior Differential Systems* (Springer, New York, 1991)
27. R.L. Bryant, S.M. Salamon, On the construction of some complete metrics with exceptional holonomy. Duke Math. J. **58**, 829–850 (1989)
28. E. Calabi, Isometric families of Kähler structures, in *The Chern Symposium* 1979, ed. by W.Y. Hsiang, et al. (Springer, New York, 1980)
29. E. Calabi, Métriques kähleriennes et fibrés holomorphes. Ann. Sci. E.N.S. 12, 269–294 (1979)
30. A. Cannas da Silva, *Lectures on Symplectic Geometry*, vol. 1764, LNM (Springer, Heidelberg, 2008). (Corrected 2nd Printing)
31. F. Cantrijn, A. Ibort, Introduction to Poisson supermanifolds. Differ. Geom. Appl. **1**, 133–152 (1991)
32. J.F. Cariñena, A. Ibort, G. Marmo, G. Morandi, *Geometry from Dynamics, Classical and Quantum* (Springer, Berlin, 2015)
33. J.F. Cariñena, G. Marmo, J. Nasarre, The non linear superposition principle and the Wei-Norman method. Int. J. Mod. Phys. A **13**, 3601–3627 (1998)
34. E. Cartan, Les groupes d'holonomie des espaces généralisés. Acta Math. **48**, 1–42 (1926)
35. E. Cartan, *Les systémes différentielles extérieurs et leur application géométriques* (Hermann, Paris, 1945, 1971)
36. P.Y. Casteill, E. Ivanov, G. Valent, Quaternionic extension of the double Taub-NUT metric. Phys. Lett. B **508**, 354–364 (2001)
37. S. Cecotti, S. Ferrara, L. Girardello, Geometry of type II superstrings and the moduli spaces of superconformal theories. Int. J. Mod. Phys. A **4**, 2475–2529 (1989)
38. K.T. Chen, Decomposition of differential equations. Math. Ann. **146**, 263–278 (1962)
39. S.A. Cherkis, Moduli spaces of instantons on the Taub-NUT space. Commun. Math. Phys. **290**, 719–736 (2009)
40. S.A. Cherkis, A. Kapustin, Hyper-Kähler metrics from periodic monopoles. Phys. Rev. D **65**, 084015 (2002)
41. D. Cherney, E. Latini, A. Waldron, Quaternionic Kahler detour complexes and $\mathcal{N} = 2$ supersymmetric black holes. Commun. Math. Phys. **302**, 843–873 (2011)
42. S.S. Chern, *Complex Manifolds Without Potential Theory* (Springer, New York, 1979)
43. S.S. Chern, W.H. Chen, K.S. Lam, *Lectures on Differential Geometry* (World Scientific, Singapore, 2000)
44. G. Cicogna, G. Gaeta, Spontaneous linearization and periodic solutions in Hopf and symmetric bifurcations. Phys. Lett. A **116**, 303–306 (1986)
45. G. Cicogna, G. Gaeta, Quaternionic-like bifurcation in the absence of symmetry. J. Phys. A **20**, 79–89 (1987)
46. G. Cicogna, G. Gaeta, *Symmetry and Perturbation Theory in Nonlinear Dynamics*, vol. 57 (Lecture Notes in Physics Monographs (Springer, Berlin, 1999)
47. G. Cicogna, S. Walcher, Convergence of normal form transformations: the role of symmetries. Acta Appl. Math. **70**, 95–111 (2002)
48. M. Cini, B. Touschek, The relativistic limit of spin 1/2 particles. Nuovo Cimento **7**, 422–423 (1958)

49. A.S. Dancer, Nahm's equations and hyperkähler geometry. Commun. Math. Phys. **158**, 545–568 (1993)
50. P.A.M. Dirac, *The Principles of Quantum Mechanics* (Pergamon, London, 1958)
51. J.J. Duistermaat, On global action-angle coordinates. Commun. Pure Appl. Math. **33**, 687–706 (1980)
52. M. Dunajski, *Solitons, Instantons and Twistors* (Oxford University Press, Oxford, 2010)
53. E.B. Dynkin, The maximal subgroups of the classical groups. Trudy Moskov Mat. Obsc. 1, 39 (1952). (English translation. In: AMS Transl. (Series 2) 6, 245 (1957))
54. T. Eguchi, P.B. Gilkey, A.J. Hanson, Gravitation, gauge theories and differential geometry. Phys. Rep. **66**, 213–393 (1980)
55. C. Elphick et al., A simple global characterization of normal forms of singular vector fields, Phys. D 29, 95–127 (1987). (Addendum, ibidem 32, 488 (1988))
56. M. Fecko, *Differential Geometry and Lie Groups for Physicists* (Cambridge UP, Cambridge, 2006)
57. L.L. Foldy, S.A. Wouthuysen, On the Dirac theory of spin 1/2 particles and its non-relativistic limit. Phys. Rev. **78**, 29–36 (1950)
58. G. Gaeta, Asymptotic symmetries and asymptotically symmetric solutions of partial differential equations. J. Phys. A **27**, 437–451 (1994)
59. G. Gaeta, Quaternionic integrability. J. Nonlinear Math. Phys. **18**, 461–474 (2011)
60. G. Gaeta, P. Morando, Hyperhamiltonian dynamics. J. Phys. A **35**, 3925–3943 (2002)
61. G. Gaeta, P. Morando, Quaternionic integrable systems, in SPT2002–Symmetry and Perturbation Theory (Proceedings of Cala Gonone workshop, May 2002), ed. by S. Abenda, G. Gaeta, S. Walcher (World Scientific, Singapore, 2003)
62. G. Gaeta, P. Morando, A variational principle for volume-preserving dynamics. J. Nonlinear Math. Phys. **10**, 539–554 (2003)
63. G. Gaeta, P. Morando, Maximal degree variational principles and Liouville dynamics. Differ. Geom. Appl. **21**, 27–40 (2004)
64. G. Gaeta, P. Morando, Variational principles for involutive systems of vector fields. Int. J. Geom. Methods Mod. Phys. **1**, 201–232 (2004)
65. G. Gaeta, P. Morando, T. Turgut, Symmetry reduction in the variational approach to Liouville dynamics. Int. J. Geom. Methods Mod. Phys. **2**, 657–674 (2005)
66. G. Gaeta, M.A. Rodríguez, On the physical applications of hyperhamiltonian dynamics. J. Phys. A 41, 175203 (2008). (16pp)
67. G. Gaeta, M.A. Rodríguez, Hyperkahler structure of the Taub-NUT metric. J. Nonlinear Math. Phys. 19, 1250014 (2012). (10 pp)
68. G. Gaeta, M.A. Rodríguez, Canonical transformations for hyperkahler structures and hyperhamiltonian dynamics. J. Math. Phys. 55, 052901 (2014). (26pp)
69. G. Gaeta, M.A. Rodríguez, Symmetry and quaternionic integrable systems. J. Geom. Phys. **87**, 134–148 (2015)
70. G. Gaeta, M.A. Rodríguez, Structure preserving transformations in hyperkahler Euclidean spaces. J. Geom. Phys. **100**, 33–51 (2016)
71. G. Gaeta, M.A. Rodríguez, Canonical transformations for hyperhamiltonian dynamics in Euclidean spaces. J. Geom. Phys. **103**, 38–52 (2017)
72. G. Gaeta, M. Spera, Remarks on the geometric quantization of the Kepler problem. Lett. Math. Phys. **16**, 189–197 (1988)
73. G. Gallavotti, *The Elements of Mechanics*, 2nd edn. (Springer 1983), http://ipparco.roma1.infn.it/pagine/deposito/2007/elements.pdf
74. G. Gentili, S. Marchiafava, M. Pontecorvo (eds.), *Proceedings of the Meeting on Quaternionic Structures in Mathematics and Physics* (SISSA, Trieste, Italy, September 5–9, 1994) pp. 270, http://www.math.unam.mx/EMIS/proceedings/QSMP94/contents.html
75. G.W. Gibbons, P. Rychenkova, R. Goto, HyperKahler quotient construction of BPS monopole moduli spaces. Commun. Math. Phys. **186**, 581–599 (1997)
76. R. Gilmore, *Lie Groups, Lie Algebras, and some of their Applications* (Wiley, New York, 1974)

77. R. Gilmore, Baker-Campbell-Hausdorff formulas. J. Math. Phys. **15**, 2090–2092 (1974)
78. M. Giordano, G. Marmo, C. Rubano, The inverse problem in the Hamiltonian formalism: integrability of linear Hamiltonian fields. Inverse Prob. **9**, 443–467 (1993)
79. C. Godbillon, *Géometrie Différentielle et Mécanique Analitique* (Hermann, Paris, 1969)
80. T. Gramchev, S. Walcher, Normal forms of maps: formal and algebraic aspects. Acta Appl. Math. **87**, 123–146 (2005)
81. V. Guillemin, S. Sternberg, *Geometric Asymptotics* (AMS, Providence, 1990)
82. V. Guillemin, S. Sternberg, *Symplectic Techniques in Physics* (Cambridge UP, Cambridge, 1984)
83. G.S. Hall, D.P. Lonie, Holonomy groups and spacetimes. Class. Quantum Gravity **17**, 1369–1382 (2000)
84. M. Hamermesh, *Group theory and its application to physical problems (Addison-Wesley, Reading, 1958)* (Dover, New York, 1989)
85. G. Heckman, *Symplectic Geometry* (Nijmegen, 2014), http://www.math.ru.nl/heckman/symplgeom.pdf
86. R. Hermann, *Differential Geometry in the Calculus of Variations* (Academic Press, New York, 1968)
87. N. Hitchin, Hyperkähler manifolds. Séminaire N. Bourbaki **748**, 137–166 (1991)
88. N. Hitchin, The Dirac operator, in Invitation to Geometry and Topology, ed. by M.R. Bridson, S.M. Salamon (Oxford UP, Oxford, 2002)
89. N. Hitchin, Generalized Calabi-Yau manifolds. Q. J. Math. **54**, 281–308 (2003)
90. N. Hitchin, Instantons, Poisson structures and generalized Kähler geometry. Commun. Math. Phys. **265**, 131–164 (2006)
91. N. Hitchin, On the Hyperkähler/Quaternion Kähler Correspondence. Commun. Math. Phys. **324**, 77–106 (2013)
92. N.J. Hitchin, A. Karlhede, U. Lindström, M. Roček, Hyperkähler metrics and Supersymmetry. Commun. Math. Phys. **108**, 535–589 (1987)
93. J.E. Humphreays, *Introduction to Lie algebras and Representation Theory* (Springer, New York, 1972)
94. D. Husemoller, *Fibre Bundles* (Springer, Berlin, 1966)
95. D. Huybrechts, *Complex Geometry: An Introduction, Universitext* (Springer, Berlin, 2005)
96. D. Huybrechts, Compact hyperkähler manifolds, in *Calabi-Yau Manifolds and Related Geometries*, ed. by G. Ellingsrud, K. Ranestad, L. Olson, S.A. Stromme (Springer, Berlin, 2002)
97. G. Iooss, M. Adelmeyer, *Topics in Bifurcation Theory and Applications* (World Scientific, Singapore, 1992)
98. C. Itzykson, J.B. Zuber, *Quantum Field Theory* (McGraw-Hill, London 1985). (reprinted by Dover, Mineola, 2006)
99. I.T. Ivanov, M. Roček, Supersymmetric σ-models, twistors, and the Atiyah-Hitchin metric. Commun. Math. Phys. **182**, 291–302 (1996)
100. D.D. Joyce, *Compact Manifolds with Special Holonomy* (Oxford UP, Oxford, 2000)
101. D. Joyce, The hypercomplex quotient and the quaternionic quotient. Math. Ann. **290**, 323–340 (1991)
102. D. Joyce, Compact 8-manifolds with holonomy Spin(7). Invent. Math. **123**, 507–552 (1996)
103. D. Joyce, Riemannian holonomy groups and calibrated geometry, in *Calabi-Yau Manifolds and Related Geometries* ed. by G. Ellingsrud, K. Ranestad, L. Olson, S.A. Stromme (Springer, Berlin, 2002)
104. D. Joyce, *Riemannian Holonomy Groups and Calibrated Geometry* (Oxford UP, Oxford, 2007)
105. A.A. Kirillov, *Elements of the Theory of Representations* (Springer, Berlin, 1976)
106. S. Kobayashi, K. Nomizu, *Foundations of Differential Geometry*, vol. I (Wiley, New York, 1963)
107. B. Kostant, Holonomy and the Lie algebra of infinitesimal motions of a Riemannian manifold. Trans. A.M.S. 80, 528–542 (1955)

108. P.B. Kronheimer, The construction of ALE spaces as hyper-Kähler quotients. J. Differ. Geom. **29**, 665–683 (1989)
109. L.D. Landau, E.M. Lifshitz, *Mechanics* (Pergamon Press, 1959)
110. L.D. Landau, E.M. Lifshitz, *Quantum Mechanics* (Pergamon Press, London, 1958)
111. C. Lanczos, *The Variational Principles of Mechanics* (Dover, New York, 1970)
112. A. Lichnerowicz, *Théorie Globale des Connexions et des Groupes d'Holonomie* (Cremonese, Roma, 1955)
113. A. Lichnerowicz, Les variétés de Poisson et leurs algèbres de Lie associées. J. Differ. Geom. **12**, 253–300 (1977)
114. U. Lindström, M. Roček, Properties of hyperkahler manifolds and their twistor spaces. Commun. Math. Phys. **293**, 257–278 (2010)
115. S. Marchiafava, P. Piccinni, M. Pontecorvo, *Quaternionic Structures in Mathematics and Physics* (World Scientific, Singapore, 2001)
116. D. McDuff, Examples of simply connected symplectic non-Kähler manifolds. J. Differ. Geom. **20**, 594–597 (1961)
117. A. Messiah, *Quantum Mechanics* (North Holland, Amsterdam, 1961)
118. J.G. Miller, M.D. Kruskal, B.B. Godfrey, Taub-NUT (Newman, Unti, Tamburino) metric and incompatible extensions. Phys. Rev. D. **4**, 2945–2948 (1971)
119. C. Misner, The flatter regions of Newman, Unti, and Tamburino's generalized Schwarzschild space. J. Math. Phys. **4**, 924–938 (1963)
120. P. Morando, M. Tarallo, Quaternionic Hamilton equations. Mod. Phys. Lett. A **18**, 1841–1847 (2003)
121. A. Moroianu, *Lectures on Kähler Geometry* (Cambridge UP, Cambridge, 2007)
122. B. Mulligan, Mass, energy, and the electron. Ann. Phys. 321, 1865–1891 (2006). (NY)
123. M. Nakahara, *Geometry, Topology and Physics* (IOP, Bristol, 1990)
124. Ch. Nash, S. Sen, *Topology and Geometry for Physicists* (Academic Press, London, 1983). (reprinted by Dover, Mineola, 2011)
125. A. Newlander, L. Nirenberg, Complex analytic coordinates in almost complex manifolds. Ann. Math. **65**, 391–404 (1957)
126. E. Newman, L. Tamburino, T. Unti, Empty-space generalization of the Schwarzschild metric. J. Math. Phys. **4**, 915–924 (1963)
127. A. Nijenhuis, On the holonomy group of linear connections I. General properties of affine connections. Indag. Math. **15**, 233–249 (1953)
128. A. Nijenhuis, On the holonomy group of linear connections II. Properties of general linear connections. Indag. Math. **16**, 17–25 (1954)
129. T. Noda, A special Lagrangian fibration in the Taub-NUT space. J. Math. Soc. Japan **60**, 653–663 (2008)
130. H. Pedersen, Y.S. Poon, Hyper-Kähler metrics and a generalization of the Bogomolny equations. Commun. Math. Phys. **117**, 569–580 (1988)
131. Y.S. Poon, S.M. Salamon, Quaternionic Kahler 8-manifolds with positive scalar curvature. J. Differ. Geom. **33**, 363–378 (1991)
132. S. Salamon, Quaternionic Kähler manifolds. Invent. Math. **67**, 143–171 (1982)
133. S. Salamon, *Riemannian Geometry and holonomy groups* (Longman, Harlow, 1989)
134. L.J. Schwachhöfer, Holonomy Groups and Algebras, in *Global Differential Geometry* ed. by C. Bär et al. (Springer, Berlin, 2012)
135. J. Simons, On the transitivity of holonomy systems. Ann. Math. **76**, 213–234 (1962)
136. A.J. Sommese, Quaternionic manifolds. Math. Ann. **212**, 191–214 (1975)
137. S. Steinberg, Applications of the Lie algebraic formulas of Baker, Campbell, Hausdorff, and Zassenhaus to the calculation of explicit solutions of partial differential equations. J. Differ. Eqs. **46**, 404–434 (1977)
138. R.S. Strichartz, The Campbell-Baker-Hausdorff-Dynkin formula and solutions of differential equations. J. Funct. Anal. **72**, 320–345 (1987)
139. A. Sudbery, Quaternionic analysis. Math. Proc. Camb. Phil. Soc. **85**, 199–225 (1979)
140. A. Swann, HyperKähler and quaternionic Kähler geometry. Math. Ann. **289**, 421–450 (1991)

141. A.H. Taub, Empty space-times admitting a three parameter group of motions. Ann. Math. **53**, 472–490 (1951)
142. D. Tong, *Lectures on Classical Dynamics*, http://www.damtp.cam.ac.uk/user/tong/dynamics.html
143. S. Walcher, *Algebras and Differential Equations* (Hadronic Press, Palm Harbor, 1991)
144. S. Walcher, On sums of vector fields. Res. Math. **31**, 161–169 (1997)
145. J. Wei, E. Norman, Lie algebraic solution of linear differential equations. J. Math. Phys. **4**, 575–581 (1963)
146. J. Wei, E. Norman, On global representation of the solutions of linear differential equations as a product of exponentials. Mem. A.M.S. 15, 327–334 (1964)
147. A. Weinstein, Lagrangian submanifolds and Hamiltonian systems. Ann. Math. **98**, 377–410 (1973)
148. A. Weinstein, Poisson geometry. Differ. Geom. Appl. **9**, 213–238 (1998)
149. E. Witten, Branes, instantons, and Taub-NUT spaces. J. High Energy Phys. **6**, 067 (2009)
150. R. Zucchini, The quaternionic geometry of four-dimensional conformal field theory. J. Geom. Phys. 27, 113–153 (1998)

Index

© Springer International Publishing AG 2017
G. Gaeta and M.A. Rodríguez, *Lectures on Hyperhamiltonian Dynamics and Physical Applications*, Mathematical Physics Studies, DOI 10.1007/978-3-319-54358-1

Printed in the United States
By Bookmasters